Chemistry 105

Lecture Notes

Dana S. Chatellier

University of Delaware

KENDALL/HUNT PUBLISHING COMPANY
4050 Westmark Drive Dubuque, Iowa 52002

ISBN13: 978-0-7575- 5177-2

Printed in the United States of America
10 9 8 7 6 5 4 3

DEDICATION

To my wife, Michelle, with all my love.

--Dana S. Chatellier
March, 2008

TABLE OF CONTENTS

PREFACE

My first teaching job was at Willamette University in Salem, Oregon, in 1985. At that time, I compiled lecture notes in handwritten form for the general chemistry classes I taught at that institution. Upon relocating to the University of Delaware in 1986, I continued to use the same lecture notes to teach first CHEM-101 and then CHEM-102. At the suggestion of a representative from the Kendall/Hunt Publishing Company, in 1989 I converted the handwritten notes into a typewritten version, which Kendall/Hunt has been publishing ever since for my CHEM-101 and CHEM-102 classes at Delaware.

A lot has happened since 1989. I married Michelle, a former student of mine, in 1990. Among the many joys of the first seventeen years of our marriage has been the opportunity to work together on numerous projects related to chemistry education. We got Chelsea, our beloved English Springer Spaniel, in 1993, and had fourteen wonderful years to enjoy with her. I taught my 25th university chemistry class in 1991. In 2008, that total will reach 225 – and counting. It's been a lot of fun!

The turn of the century has brought impressive technological innovations which were unknown when I began my teaching career. We now have cell phones, the Internet, iPods, Tivo, and so much more. I'm typing this preface using Microsoft Word (although I still type exams on the old Smith-Corona typewriter at home – old habits die hard, but at least it's an electric typewriter, not the manual typewriter I used while I was at Willamette). My students today, who were infants (or unborn) when I first came to Delaware, have grown up with these tools and expect to use them to further their education, as they should. And speaking of my students today....

The fall of 2007 marked my first attempt to teach CHEM-105, a one-semester course in general chemistry primarily intended for nursing majors. My notes for the class were essentially a compilation of numerous topics taken from my CHEM-101 and CHEM-102 lecture notes, used verbatim. What you hold in your hand is Kendall/Hunt's formal publication of these notes, especially designed for CHEM-105 students. This book is not a textbook, but is intended to supplement the textbook for the class. It is written in a style that is more conversational than the language found in most textbooks – in short, it's written the way I teach it! Many of my students have found the lecture notes to be very beneficial.

Chemistry is a fascinating field of study, and one whose importance in the 21st century cannot be denied. ("Biotechnology", "genetic engineering", "materials science", and "nanotechnology" are all "chemistry", in one form or another.) If this book helps you to understand the world and the age in which you live, then it will have served its purpose. I hope you find it helpful.

--Dana S. Chatellier
March, 2008

TO THE STUDENT:

This is *your* book. You've bought it and paid for it -- it's yours. Use it that way! Make notes to yourself in the margins or between lines when you feel the need. (That's why it was typed double-spaced!) Draw pictures to illustrate an example or a chemical demonstration done in class. (This book is almost pictureless, since there are many good illustrations in your textbook.) Tear out pages and rearrange them when necessary. Throw the book at the wall (not at your roommate!) when studying chemistry gets you frustrated! This is *your* book, so use it however *you* see fit.

Good luck! The study of chemistry is challenging, but most students find that they can master it with enough hard work and patience. Read the textbook and these lecture notes, work practice problems, and see your instructor if you're having problems. (Most students don't admit that they're having problems until they do poorly on a few exams. That's too late! See your instructors *before* you run into serious problems -- it will make for a much more enjoyable experience for everyone.)

Best wishes for a rewarding semester learning about chemistry.

--Dana S. Chatellier
February, 1989

SCIENCE AND THE SCIENTIFIC METHOD:

Science is a way of trying to understand the universe in which we live. The key word here is "understand". In the words of author Robert Heinlein, "The difference between science and the fuzzy subjects is that science requires reasoning, while those other subjects merely require scholarship."[1]

Scientists, according to mathematician John von Neumann, "mainly make models. A model is a mathematical construct which describes observed phenomena. The justification of such a construct is that it is expected to work."[2] Adds Leo Kadanoff, "One is surprised that a construct of one's own mind can actually be realized in the honest-to-goodness world out there. A great shock, a great joy."[2]

von Neumann's comments are a brief way of summarizing the scientific method, which is the scientist's way of trying to understand the universe. The scientific method involves choosing a natural phenomenon to be studied, doing experiments to see how the object being studied interacts with other things, and recording the results of these experiments as data. Data may include both qualitative observations and quantitative measurements. If any patterns or trends are observed in a collection of data, and if further experiments confirm that these trends are broadly applicable, then the trends are referred to as laws. Laws tell what happens, but they say nothing about why it happens. Scientists form models called hypotheses to try to explain why things happen. A hypothesis whose predictions are confirmed over a long period of time is called a theory. Theories and laws are never proven, but they can be disproven. "No amount of experimentation can ever prove me right; a single experiment can prove me wrong." --A. Einstein[3]

---

1) Heinlein, R. A., Time Enough For Love. G. P. Putnam's Sons, 1973.

2) Quoted by James Gleick in Chaos: Making a New Science. Penguin Books, 1987.

3) Quoted by Dr. Laurence J. Peter in Peter's Quotations. William Morrow and Company, Inc., 1977.

CHEMISTRY -- THE CENTRAL SCIENCE:

In the words of Dr. John A. Conkling, Executive Director of the American Pyrotechnics Association, "Chemistry is a science concerned with substances and their properties, with the changes or reactions whereby other substances are formed, with the conditions necessary for bringing about or preventing these changes, and with the relative amounts of matter and energy involved."[1] Chemistry has been called "the central science" because it is so closely related to other sciences. For example, among the many branches of chemistry are biochemistry (the chemistry of living things), organic chemistry (the chemistry of the substances of which living things are made), and physical chemistry (chemistry related to physics, involving studies of matter and energy).

Matter is the "stuff" of which any physical object is made -- anything that has mass (or weight) and occupies space is made of matter. Matter has many properties (or characteristics) which can be measured. The properties of matter can be classified as either extensive properties (which depend on the amount of matter present) or intensive properties (which don't depend on the amount of matter present). Extensive properties include the mass and volume of an object; intensive properties include the object's melting point and boiling point. Note that an object's density (an intensive property) is just the ratio of its mass to its volume (two extensive properties): **Density = mass/volume.**

The properties of matter can also be classified as chemical properties (which depend on the substance's interaction with other substances) or as physical properties (which don't depend on the substance's interaction with any other substance). The properties mentioned above are all physical properties. Examples of chemical properties include a substance's acidity, flammability, and toxicity.

---

1) From a lecture given at Washington College, Chestertown, Maryland, September, 1974.

CLASSIFICATIONS OF MATTER:

Most of the matter which is found in nature is a mixture of two or more substances. Impure substances have variable compositions -- that is, they can be mixed in any proportion that we choose. As an example, consider coffee -- a mixture of water, caffeine, sugar, cream, etc. These ingredients can be mixed in whatever proportions are desired. This is fortunate, since some people like their coffee stronger (or sweeter, or blacker) than others.

Impure substances occur in two varieties: they are either homogeneous mixtures or heterogeneous mixtures. Heterogeneous mixtures are composed of at least two distinct phases. Phases are regions whose physical and chemical properties differ from each other, but are constant within that region. As an example, consider a mixture of oil and water. "Oil and water don't mix" -- they separate into two layers: an oil layer (oil phase) and a water layer (water phase). Mixtures which contain only one phase are called homogeneous mixtures. As an example, consider the cup of coffee again. Coffee, if stirred properly, does not separate into a water layer, a cream layer, a caffeine layer, and so on. There's just one phase (the "coffee phase") in a homogeneous mixture.

Much of the hard work that many chemists do involves separating mixtures into pure substances, which have constant compositions. Pure substances fall into one of two classifications: elements and compounds. Elements are the simplest forms of matter. They cannot be broken down into simpler substances by normal chemical methods. The smallest particles into which an element can be divided without losing its chemical properties are called atoms. Each element is composed of only one kind of atom. Compounds are formed from two or more elements combined chemically in a fixed ratio. Many compounds are made of molecules, which are small particles composed of two or more atoms.

THE CHEMIST'S "SHORTHAND":

When chemists write about the elements and compounds that they use, they often use a "shorthand" notation to be able to avoid writing the full names of the substances involved. Learn this "shorthand" notation -- it saves time!

Elements are written using chemical symbols. Chemical symbols are simply one-letter or two-letter abbreviations for the names of the elements. One-letter symbols are used for many common elements, such as hydrogen (H), carbon (C), and nitrogen (N). A second letter is added to distinguish between two elements whose names begin with the same letter. Elements with two-letter symbols include helium (He), chlorine (Cl), and niobium (Nb). A few elements have symbols which are derived from their Latin names. Examples include mercury (Hg, from the Latin hydrargyrum), copper (Cu, from cuprum), and sodium (Na, from natrium). You should learn the symbols and names of the more common elements as soon as you can.

Compounds are written using chemical formulas. Water is a compound whose formula is the well-known H_2O. This formula means that one molecule of water is made of two atoms of hydrogen (H) and one atom of oxygen (O) which are held together in some way. Many compounds are made of more complex molecules. A common example is sucrose (table sugar), which has the formula $C_{12}H_{22}O_{11}$. This formula indicates that twelve carbon atoms, twenty-two hydrogen atoms, and eleven oxygen atoms are joined to each other to form one molecule of sucrose.

Hydrates are compounds whose crystals contain loosely-held molecules of water in their crystal structures. Their formulas are written so as to clearly indicate this. For example, the formula $CoCl_2 \cdot 6H_2O$ shows that six molecules of water are connected to each $CoCl_2$ unit. (Note the importance of the correct use of capital letters when writing chemical symbols and formulas. "$CoCl_2$" is a crystalline solid; "$COCl_2$" is a poisonous gas. Little things mean a lot!)

kilo - 1000
centi - 100
mili -

THE METRIC SYSTEM:

By international agreement, scientists use the metric system to measure the various quantities that they need to know in their laboratories. The metric system is also called the SI system. SI comes from Systeme Internationale d'Unites -- the International System of Units. Most industrialized countries (with the notable exception of the United States!) use the metric system for everyday transactions.

One of the advantages of the metric system is that it is based on powers of ten, which makes it easy to convert from one unit to another. In the metric system, certain prefixes are used to indicate multiples or fractions of the base unit. For example, "kilo-" means one thousand of the base unit (1000 grams = 1 kilogram), "centi-" means one one-hundredth of the base unit (100 centimeters = 1 meter), and "milli-" means one one-thousandth of the base unit (1000 milliliters = 1 liter). By contrast, there are two pints in a quart, three feet in a yard, four quarts in a gallon, twelve inches in a foot, etc. Use metric -- it's easier!

Conversion to the metric system has met with resistance in the United States, presumably because people are unfamiliar with the ways in which metric units compare with other units. A table of metric conversion factors is provided.

remember the basic conversions

Measured Quantity	SI Unit of Measure	Conversion Factor Equations
Mass	Kilogram	1 kg = 2.205 pounds
	Gram	28.35 g = 1 ounce
Distance	Meter	1 m = 39.37 inches
	Kilometer	1.609 km = 1 mile
Volume	Cubic Meter	1 m^3 = 1000 liters
	Liter	1 L = 1.057 quarts

MEASURING TEMPERATURE:

In the United States, temperature is measured using the Fahrenheit scale, but in most other countries, the Celsius (sometimes called centigrade) scale is used. Scientists use the Celsius scale and another scale called the Kelvin scale in the laboratory. These two temperature scales are well-suited for the kinds of measurements and calculations that scientists typically make.

On the Fahrenheit scale, the freezing point of water is 32 ^{o}F, and the boiling point of water is 212 ^{o}F. The Celsius scale is sometimes called the centigrade scale because these same two physical phenomena, the freezing and boiling of water, are separated by one hundred degrees ("centi-" means 1/100.). The freezing point of water is 0 ^{o}C, and the boiling point of water is 100 ^{o}C. The same two points are also separated by one hundred Kelvins (Note: not "degrees Kelvin"!): water freezes at 273 K, and water boils at 373 K. Zero on the Kelvin scale corresponds to absolute zero, the coldest possible temperature. This equals -459 ^{o}F and -273 ^{o}C, but there are no "below zero" temperatures on the Kelvin scale. The Kelvin scale is sometimes called the absolute scale for this reason.

Converting a temperature from Celsius to Kelvin is easy: just add 273! Similarly, just subtract 273 from a Kelvin temperature to get the corresponding Celsius temperature. Conversion between the Fahrenheit and Celsius scales is a little bit trickier, but here's a relatively simple way to do it:[1] Step 1 -- add 40. Step 2 -- multiply by 9/5 if converting Celsius to Fahrenheit. Multiply by 5/9 if converting Fahrenheit to Celsius. Step 3 -- subtract 40.

Summary: $K = {}^{o}C + 273.$ $\quad {}^{o}F = [(40 + {}^{o}C)(9/5)] - 40 = (1.8)({}^{o}C) + 32$

$${}^{o}C = K - 273. \quad {}^{o}C = [(40 + {}^{o}F)(5/9)] - 40 = \frac{({}^{o}F - 32)}{1.8}.$$

1) The author wishes to thank Dr. John Burmeister, University of Delaware, for sharing this method.

USING CONVERSION FACTORS: to convert feet into yards divide by 3

To convert 300 feet into yards, we need only to divide by 3, since 3 feet = 1 yard. (Everyone knows that!) But how would you approach such a problem if you didn't know simply to divide by 3 -- that is, if you were using units which aren't as familiar as feet and yards?

A conversion factor is simply an algebraic expression that equals one. As an example, let's consider the equation above: 3 feet = 1 yard. Since this is a true statement, then the following two equations must also be true:

$$\frac{3\text{ feet}}{1\text{ yard}} = 1 \qquad\qquad \frac{1\text{ yard}}{3\text{ feet}} = 1$$

To convert 300 feet into yards, all we need to do is multiply by one in the form of one of the above conversion factors. But which one shall we use? If we choose the conversion factor on the left (above), we get this result:

$$300\text{ feet} \times \frac{3\text{ feet}}{1\text{ yard}} = 900\ \frac{\text{feet x feet}}{\text{yards}} = 900\text{ feet}^2/\text{yard}.$$

need to be on the bottom

This is clearly incorrect, since we know that our final answer should be in yards, not some silly units like feet^2/yard! Using the other conversion factor above gives the correct result, complete with the correct units:

$$300\text{ feet} \times \frac{1\text{ yard}}{3\text{ feet}} = 100\ \frac{\text{feet x yards}}{\text{feet}} = 100\text{ yards}.$$

The general method is to set up the conversion factors so that any undesired units will appear in both the numerator and the denominator of the final expression. These units will cancel out, leaving only the desired units.

Problem: A European car gets 12.5 kilometers per liter of gasoline. What is the mileage in miles per gallon? 1 mile = 1.609 km. 1 gallon = 3.786 L.

Solution: Setting up the necessary conversion factors, we get:

$$12.5\ \frac{\text{km}}{\text{L}} \times \frac{3.786\text{ L}}{1\text{ gallon}} \times \frac{1\text{ mile}}{1.609\text{ km}} = 29.4\ \frac{\text{miles}}{\text{gallon}}.$$

SCIENTIFIC NOTATION:

Problem: Light travels through space at a speed of 300,000,000 meters per second. Express this velocity in furlongs per fortnight!

Conversion Equations: 8 furlongs = 1 mile. 14 days = 1 fortnight.

Solution: Set up the necessary conversion factors!

$$300{,}000{,}000\ \frac{\text{m}}{\text{sec}} \times \frac{60\ \text{sec}}{1\ \text{min}} \times \frac{60\ \text{min}}{1\ \text{hr}} \times \frac{24\ \text{hr}}{1\ \text{day}} \times \frac{14\ \text{days}}{1\ \text{fortnight}}$$

$$\times \frac{1\ \text{km}}{1000\ \text{m}} \times \frac{1\ \text{mile}}{1.609\ \text{km}} \times \frac{8\ \text{furlongs}}{1\ \text{mile}} = 1{,}800{,}000{,}000{,}000\ \frac{\text{furlongs}}{\text{fortnight}}.$$

Scientists often deal with large numbers such as 300,000,000 and 1,800,000,000,000. They also need to use small numbers such as 0.000000656. Scientific notation has been developed to enable scientists to work with very large and very small numbers conveniently (and to save a lot of time and effort which would otherwise be spent writing a lot of zeroes!). To write a number in scientific notation, simply move the decimal point until the number that is formed is between one (1) and ten (10). Then, write "$\times 10^n$" after the new number. The value of n is simply the number of decimal places that the decimal point would have to be moved to return to the original number. Some examples are provided.

Problem: Write 300,000,000 in scientific notation.

Solution: Another way to write 300,000,000 is 300000000. Notice that the decimal point occurs after the last zero. Moving this decimal point eight places to the left gives us the number 3. Therefore, the value of n is 8, and 300,000,000 would be written as 3×10^8 in scientific notation.

Problem: Write 0.000000656 in scientific notation.

Solution: Moving the decimal point seven places to the right gives us a number between one and ten -- namely, 6.56. Therefore, the value of n is -7, and 0.000000656 would be written as 6.56×10^{-7} in scientific notation.

SIGNIFICANT FIGURES:

All measurements contain uncertainties. Surveys and opinion polls are usually reported as having a certain "margin for error". When you measure something (using a ruler or thermometer, for example), you can only read the number to a certain degree of precision -- your eyes work well enough to allow you to "read between the lines" to a certain extent, but not much more than that.

When a scientist measures something, s/he records not only the numerical value of the measurement, but also an indication of the uncertainty of the measurement. Suppose, for example, a mass was determined using a balance which can be read to the nearest centigram (0.01 gram). If the measured mass was 6.03 grams, the entry in the scientist's laboratory notebook would look something like this: "Mass of liquid = 6.03 ± 0.01 grams". (The "±" symbol is read as "plus or minus", implying an uncertainty of 0.01 grams.) For the scientist to record any digits beyond the "3" in the hundredths place would be meaningless, since there is an uncertainty in that place. If the hundredths place contains an uncertainty, the thousandths place and others beyond it must also be uncertain.

If the mass of the liquid had been recorded as simply "6.03 grams", an uncertainty in the hundredths place would be assumed. This is referred to as the significant figure convention -- the significant digits are all of those which are known with certainty, plus one digit which is uncertain. In the above case, the "6" and the "0" are known with certainty, and the "3" is uncertain. Hence, the measured number "6.03" is said to have three significant figures ("sig figs").

If you are not sure how many "sig figs" are present in a measured number, write the number in scientific notation! For example, the number 0.082060 may appear to have seven "sig figs", but writing it as 8.2060×10^{-2} shows that only five "sig figs" are present -- the first two zeroes are just "place holders".

USING SIGNIFICANT FIGURES IN CALCULATIONS:

Problem: If a sample of a liquid has a mass of 6.03 grams and a volume of 7.12 milliliters, what is the density of the liquid?

Solution: Since density = mass/volume, a calculator gives us:

$$\text{Density of Liquid} = \frac{\text{Mass of Liquid}}{\text{Volume of Liquid}} = \frac{6.03\ \text{g}}{7.12\ \text{mL}} = 0.8469101\ \text{g/mL}.$$

However, it is not possible for the density to be known to seven significant figures when the measurements from which it is calculated are known to only three significant figures. (Calculators almost always give the wrong number of significant figures -- be aware of this!) At what point should we "round off"?

Since we know that the "3" in "6.03 g" and the "2" in "7.12 mL" are the uncertain digits, let's see what effect a small change in each of these digits has on the calculated density.

$$\frac{6.02\ \text{g}}{7.13\ \text{mL}} = 0.8443197\ \text{g/mL} \qquad \frac{6.04\ \text{g}}{7.11\ \text{mL}} = 0.8495077\ \text{g/mL}$$

Notice that for each of the three calculated densities above, the "0.84" at the beginning remains the same. The first digit which differs (that is, the first uncertain digit) is the digit in the thousandths place -- the third significant digit. (The zero to the left of the decimal point is not significant. Write the number in scientific notation if you're not sure of this!) Therefore, according to the significant figure convention (that is, record all of the certain digits plus one uncertain digit), the "rounding off" should occur after the thousandths place, and the density should be recorded as "0.847 g/mL". Note that three significant digits are present in the measured mass, the measured volume, and the calculated density. This is typical for calculations involving only multiplication and division -- the calculated number has the same number of significant digits as the measured numbers do, if their "sig figs" are the same.

USING SIGNIFICANT FIGURES IN CALCULATIONS (ctd.):

To be sure that a number which has been calculated from measured numbers has the correct number of significant figures, follow the rules below.

A. Multiplication and Division

When multiplying or dividing two numbers, the number of significant figures in the result is the same as the number of significant figures in the measured number which has the least number of significant figures. Example:

$$\frac{6.03\text{ g}}{7.1\text{ mL}} = 0.8492957 \text{ (calculator)} = 0.85 \text{ (correctly rounded)}$$

Since "6.03" contains three significant digits but "7.1" only has two, the result of the division should be a number with only two significant digits.

Problem: What is 6.03 m x 7.1 m ? (Solution: 43 m^2. Two sig figs.)

B. Addition and Subtraction

Since addition and subtraction involve adding or subtracting columns of numbers, the result should include the digit from the leftmost column which contains an uncertain digit, but no digits after that digit. Examples:

$$\begin{array}{r} 7.1\text{ g} \\ +\ 6.03\text{ g} \\ \hline 13.1\text{ g} \end{array} \qquad \begin{array}{r} 7.1\text{ g} \\ -\ 6.03\text{ g} \\ \hline 1.1\text{ g} \end{array}$$

In each case, the tenths column contains an uncertain digit (the "1" in "7.1"). Hence, no digits past the tenths place are written.

C. Using Defined Numbers

Problem: Express 365.24 feet in yards.

Solution: Since 3 feet = 1 yard by definition (not by measurement!), the "3" in the conversion factor "3 feet/yard" is considered to contain an infinite number of significant figures. (If you like, "3.0000000000...feet/yard!) Therefore, since "365.24 feet" contains five significant figures, the final result should also contain five significant figures. Hence:

$$365.24 \text{ feet} \times \frac{1\text{ yard}}{3\text{ feet}} = 121.74666 \text{ (calculator)} = \underline{121.75\text{ yards}}.$$

ANTOINE LAVOISIER'S COMBUSTION EXPERIMENTS:

The French scientist Antoine Lavoisier has been called the "father of modern chemistry" for his late 18th-century combustion experiments. Combustion is simply the reaction of something with oxygen. This is what happens when something catches fire and burns. Lavoisier was the first person to carefully measure the masses of the materials used and products formed in combustion reactions. Data similar to those obtained by Lavoisier during these experiments are given below.

Expt. No.	Mass of N_2 Used	Mass of O_2 Used	Product and its Mass
1	28.0 g	32.0 g	60.0 g of NO
2	28.0 g	64.0 g	92.0 g of NO_2
3	56.0 g	32.0 g	88.0 g of N_2O

laws

The results of the experiments above and many other experiments like them enabled Lavoisier to determine the following laws of chemical combination.

A. The Law of Conservation of Mass. In a chemical reaction, the total mass present remains constant -- matter is neither created nor destroyed. For example, see Experiment # 1, above. 28.0 g + 32.0 g = 60.0 g.

B. The Law of Definite Proportions. (Also called the Law of Constant Composition.) The relative amount of each element in a particular compound is always the same. For example, the compound NO contains nitrogen and oxygen in a 28:32 mass ratio. If some other ratio is used, the product is not NO!

C. The Law of Multiple Proportions. If two elements can combine so as to form more than one compound, then the masses of one element that react with a fixed mass of the other element will be related by small whole number ratios. For example, compare Experiments # 1 and # 2, above. The mass of N_2 is fixed, and the masses of O_2 which react with it are 32.0 g and 64.0 g -- a ratio of 1:2. Experiments # 1 and # 3 are similar -- a 1:2 ratio of N_2 for a fixed mass of O_2.

DALTON'S ATOMIC THEORY:

In the early part of the 19th century, the English scientist John Dalton proposed a hypothesis to explain Lavoisier's laws of chemical reactions. The main ideas of Dalton's hypothesis are listed below.

a) All elements are composed of small, indivisible, indestructible particles called atoms.

b) The atoms of any one element all have the same chemical and physical properties. Atoms of two different elements have different chemical and physical properties.

c) Molecules are formed by joining two or more atoms. Since atoms are indivisible, whole numbers of atoms must be joined in this process -- there's no such thing as a "fraction" of an atom.

d) In a chemical reaction, the atoms involved are simply rearranged to form different molecules. No atoms are created or destroyed in a reaction.

This hypothesis was sufficient to explain Lavoisier's laws. Since atoms are neither created nor destroyed during a reaction, no mass can be gained or lost (Law of Conservation of Mass). Since atoms are indivisible (that is, they cannot be cut into smaller pieces), they must combine in whole number ratios (Law of Multiple Proportions). The Law of Definite Proportions is explained similarly.

Dalton's hypothesis is a shining example of the way science usually works. Dalton's ideas about atoms and molecules are essentially a model, whose sole purpose was to explain Lavoisier's results. Dalton's hypothesis has been revised and modified in the last two centuries. Today, for example, we know that atoms can be cut into smaller pieces (nuclear fission!), and that not all atoms of an element have identical physical properties (isotopes!). But Dalton's main ideas have survived the test of time, and are called the atomic theory today.

THE DISCOVERY OF ELECTRONS:

One of the major revisions that has been made in Dalton's atomic theory is that atoms can be divided into smaller particles. These particles were discovered through a series of experiments done in the late 19th century and early 20th century.

Cathode ray tubes (CRTs) are the "picture tubes" used in television sets and computer terminals. Cathode rays are produced when a high electrical voltage is applied to a glass tube filled with a gas at low pressure. In 1897, the British physicist J. J. Thomson discovered that cathode rays are deflected by external electric and magnetic fields, are attracted to objects which have positive electrical charges, and can cause a small paddlewheel placed in their path to turn. These results allowed Thomson to conclude that cathode rays are composed of small, negatively-charged particles which he called electrons. He reasoned that these electrons were coming from the atoms of the gas in the cathode ray tube. He was also able to determine the ratio of an electron's electrical charge (e) to its mass (m): $e/m = -1.76 \times 10^{8}$ coulombs/gram. (The coulomb is the SI unit of electrical charge.) Thomson received the 1906 Nobel Prize in Physics.

In 1909, the American physicist Robert Millikan was able to determine the mass of an electron by means of his clever "oil-drop" experiment. Millikan used X-rays to ionize air molecules (that is, to make them give off electrons). The free electrons generated by this process became attached to droplets of oil that Millikan caused to fall between two electrically-charged metal plates. Millikan could make the negatively-charged oil droplets stop falling by adjusting the electrical charges on the metal plates. This enabled him to calculate the electron's electrical charge ($e = -1.60 \times 10^{-19}$ coulombs), and to use Thomson's ratio to calculate the mass of an electron ($m = 9.11 \times 10^{-28}$ grams).

PROTONS AND THE ATOMIC NUCLEUS:

Zinc sulfide glows when charged particles strike it. (This is how the glowing images on television screens are produced!) In 1898, the German physicist Wilhelm Wien found that a sheet of zinc sulfide glows when it is placed behind the negative electrode of a cathode ray tube. From this, Wien concluded that there must be positively-charged particles in cathode ray tubes, since such particles would be attracted to the negatively-charged electrode. Wien was able to measure the masses of these particles, and found that different masses were obtained when different gases were used in the cathode ray tube. The lightest particles came from hydrogen gas, and the particles obtained from other gases had masses which were approximately multiples of the mass of the hydrogen particles. Wien named the hydrogen particles protons, and received the 1911 Nobel Prize in Physics.

In 1911, the British physicist Ernest Rutherford (who had won the 1908 Nobel Prize in Chemistry for his studies of radioactivity) did an experiment in which he directed a beam of alpha particles at a thin sheet of gold foil. (Alpha particles, which are emitted by some radioactive atoms, are positively charged.) Rutherford found that while most of the alpha particles passed directly through the gold foil, a few of them were deflected from the foil at an angle. Some of them were even directed back toward their source, which to Rutherford was like "a cannonball bouncing off of tissue paper"! These results effectively disproved the "plum pudding" model of the atom which had been proposed by J. J. Thomson. This hypothesis stated that electrons were distributed randomly through through a spherical, positively-charged atom. Rutherford's results showed that atoms are mostly empty space, but contain relatively massive positively-charged particles. Rutherford proposed a new model of the atom, in which the positive charges were centered in a massive atomic nucleus, which was surrounded by the electrons.

NEUTRONS, ATOMIC NUMBERS, AND MASS NUMBERS:

After Rutherford proposed his nuclear model of the atom, he did other experiments to try to calculate the mass of the atomic nucleus. He tried to do this by measuring the charge of the atomic nucleus and calculating the mass based on Wien's values for the mass and charge of the proton, but these calculations gave results that were approximately half of the masses of the atomic nuclei that he was able to measure experimentally. Rutherford concluded that atomic nuclei contain not only protons, but also particles with about the same masses as protons but with no electrical charge. These particles were called neutrons, because they were neither positively-charged nor negatively-charged -- they were neutral. (In 1935, the British physicist James Chadwick won the Nobel Prize in Physics for his experiments which showed clearly that atomic nuclei contain neutrons.)

The table below summarizes the results of Thomson's, Millikan's, Wien's, and Rutherford's experiments. Masses are given in atomic mass units (amu) and electrical charges are based on a value of "+1.0" for the proton. The actual charge on a proton is +1.6 x 10^{-19} coulombs, and 1.0 amu = 1.66 x 10^{-24} grams.

Subatomic Particle	Mass, amu	Charge	Location in Atom
Proton	1.0	+1.0	center (nucleus)
Neutron	1.0	0.0	center (nucleus)
Electron	"0.0"	-1.0	outside the nucleus

Many physical and chemical properties of atoms are determined by the number of electrons surrounding the nucleus. This number will be the same as the number of protons in the nucleus if the atom is electrically neutral (as most atoms are). This number is called the atomic number of the atom. The mass number of an atom is simply the number of "massive" particles (protons and neutrons) its nucleus contains.

ISOTOPES, ATOMIC WEIGHTS, AND MOLECULAR WEIGHTS:

A major modification of Dalton's atomic theory involves the existence of isotopes. Isotopes are atoms of the same element (and therefore, atoms which have the same atomic number) which have different physical properties. Specifically, two isotopes differ in their masses by virtue of the fact that they contain different numbers of neutrons. Isotopes are usually represented by symbols of the general form $^{A}_{Z}Q$, where Q is the chemical symbol for the element, Z is its atomic number, and A is its mass number. For example, an atom of carbon-14 ($^{14}_{6}C$) contains six protons, six electrons, and eight (14 - 6) neutrons. Another isotope of carbon, carbon-12 ($^{12}_{6}C$), contains only six neutrons per atom.

The atomic weight of an element is the average mass of its atoms. We can speak of average masses because different isotopes occur naturally to various extents. For example, naturally-occurring chlorine is about 75% chlorine-35 ($^{35}_{17}Cl$) and 25% chlorine-37 ($^{37}_{17}Cl$). The atomic weight of chlorine is approximately 35.5 amu, since (0.75)(35 amu) + (0.25)(37 amu) = 35.5 amu. The atomic weight and atomic number of each element can be found near its symbol on the Periodic Table.

An important property of a compound is its molecular weight (sometimes called its formula weight). The molecular weight of a compound is simply the sum of the atomic weights of the atoms in its formula. As an example, the molecular weight of sodium chloride (NaCl, table salt) is simply the sum of the atomic weights of sodium (Na = 22.98977 amu) and chlorine (Cl = 35.453 amu) -- that is, 58.443 amu. The molecular weight of water (H_2O) is found by adding the atomic weight of oxygen (O = 15.9994 amu) to twice the atomic weight of hydrogen (2 x H = 2 x 1.0079 amu = 2.0158 amu) -- the result is 18.0152 amu. The molecular weight of sucrose (table sugar, $C_{12}H_{22}O_{11}$) is calculated similarly (carbon = 12.011 amu): (12 x 12.011 amu) + (22 x 1.0079 amu) + (11 x 15.9994 amu) = 342.299 amu.

THE MOLE CONCEPT:

How many socks are there in a pair of socks? Right -- two.

Slightly harder, now: How many 3¢-stamps are there in a dozen? Right -- twelve! (The same number of things there are in a dozen of anything!)

"Pair" and "dozen" are just words that represent numbers. "Pair" represents the number two; "dozen" represents the number twelve. Chemists use the word "mole" to represent a very large number -- 6.02×10^{23}. (This number is sometimes called Avogadro's Number, after the Italian scientist Amadeo Avogadro.) This number is so huge that it's useless for counting large things like socks or stamps. What it is useful for is counting small things like atoms and molecules.

The mole concept allows us to think about substances on both the large (visible) scale and the small (sub-microscopic) scale simultaneously. This is due to the fact that one mole of atoms of an element has the same mass as the atomic weight of the element, expressed in grams. For example, the element neon (Ne) has an atomic weight of 20.179 amu. Another way of saying this is to state that a sample of neon weighing 20.179 grams contains one mole (6.02×10^{23}) of Ne atoms.

Since molecular weights are simply the sums of atomic weights, it is also true that one mole of molecules of a compound has the same mass as the molecular weight of the compound, expressed in grams. For example, water (H_2O) has a molecular weight of 18.0152 amu. Therefore, one mole of water molecules has a mass of 18.0152 grams. Sucrose (table sugar, $C_{12}H_{22}O_{11}$) has a molecular weight of 342.299 amu, so Avogadro's Number of sucrose molecules weighs 342.299 grams. The molecular weight of sodium chloride (table salt, NaCl) is 58.443 amu; sodium chloride contains 58.443 grams per mole. (When referring to atomic weights or molecular weights, the units "amu" or "grams per mole" (sometimes written "g/mol") may be used interchangeably.)

USING THE MOLE CONCEPT IN CHEMICAL CALCULATIONS:

Problem: What mass of argon gas equals 0.375 moles of argon atoms?

Solution: Argon (Ar) has an atomic weight of 39.948 amu. Therefore, the conversion equation is: 39.948 grams of Ar = 1 mole of Ar atoms. Setting up the appropriate conversion factor, we obtain the following result:

$$0.375 \text{ moles Ar} \times \frac{39.948 \text{ g Ar}}{1 \text{ mole Ar}} = 15.0 \text{ g Ar.}$$

Problem: How many atoms of lead are present in 50.0 grams of lead?

Solution: Lead (Pb) has an atomic weight of 207.2 amu (207.2 g/mole). The first step is to use this conversion factor to obtain an amount in moles:

$$50.0 \text{ g Pb} \times \frac{1 \text{ mol Pb}}{207.2 \text{ g Pb}} = 0.241 \text{ moles Pb.}$$

Now, use Avogadro's Number to convert this amount into a number of atoms. The conversion equation is: 1 mole = 6.02×10^{23} atoms. Therefore:

$$0.241 \text{ moles Pb} \times \frac{6.02 \times 10^{23} \text{ Pb atoms}}{1 \text{ mole Pb}} = 1.45 \times 10^{23} \text{ Pb atoms.}$$

Problem: A sample of methane (CH_4) contains 3.75×10^{24} hydrogen atoms. Calculate the mass of the sample of methane.

Solution: Since each methane molecule contains four hydrogen atoms, we can find the number of methane molecules using the following conversion equation: 1 CH_4 molecule = 4 H atoms. Using this as a conversion factor, we get:

$$3.75 \times 10^{24} \text{ H atoms} \times \frac{1 \text{ } CH_4 \text{ molecule}}{4 \text{ H atoms}} = 9.38 \times 10^{23} \text{ } CH_4 \text{ molecules.}$$

Now, use Avogadro's Number to convert this to an amount in moles:

$$9.38 \times 10^{23} \text{ } CH_4 \text{ molecules} \times \frac{1 \text{ mole } CH_4}{6.02 \times 10^{23} \text{ molecules}} = 1.56 \text{ moles } CH_4.$$

Finally, use methane's molecular weight to convert this to a mass:

$$1.56 \text{ moles } CH_4 \times \frac{16.043 \text{ g } CH_4}{1 \text{ mole } CH_4} = 25.0 \text{ grams of } CH_4.$$

DETERMINATION OF A FORMULA FROM PERCENTAGE COMPOSITION DATA:

The percentage by mass of an element in a compound is the same as the mass (in grams) of that element needed to form 100.00 grams of the compound, except it's expressed as a percentage. As an example, consider sodium chloride: one mole (58.443 g) of NaCl is made from one mole of Na (22.98977 g) and one mole of Cl (35.453 g), so the percentages of Na and Cl in NaCl are easily calculated:

$$\% \text{ Na} = \frac{22.98977 \text{ g Na}}{58.443 \text{ g NaCl}} \times 100.00\% = 39.337\% \text{ Na by mass.}$$

$$\% \text{ Cl} = \frac{35.453 \text{ g Cl}}{58.443 \text{ g NaCl}} \times 100.00\% = 60.663\% \text{ Cl by mass.}$$

Therefore, to prepare 100.000 grams of sodium chloride, 39.337 grams of sodium and 60.663 grams of chlorine are needed. (Note that mass is conserved!)

Problem: Calculate the percentages by mass of carbon and hydrogen in the compound methane (CH_4). (Approximate Solution: 75% C, 25% H.)

Usually, the percentages by mass of the elements in a compound are determined experimentally, and then the formula of the compound is calculated from the percentage composition data.

Problem: The compound isopropyl alcohol ("rubbing alcohol") contains 60.0% C, 13.4% H, and 26.6% O by mass. What is the formula of isopropyl alcohol?

Solution: The percentages tell us that 100.0 g of alcohol is made from 60.0 g of C, 13.4 g of H, and 26.6 g of O. Convert these to moles of atoms:

$$\frac{60.0 \text{ g C atoms}}{12.011 \text{ g/mole C}} = 5.00 \text{ moles C} \qquad \frac{13.4 \text{ g H atoms}}{1.0079 \text{ g/mole H}} = 13.3 \text{ moles H}$$

$$\frac{26.6 \text{ g O atoms}}{15.9994 \text{ g/mole O}} = 1.66 \text{ moles O} \qquad \text{Formula} = C_{5.00}H_{13.3}O_{1.66} \quad (???)$$

Dalton's atomic theory says that we can't have .3 or .66 of an atom, so we need whole numbers with a 5.00/13.3/1.66 ratio. Dividing each subscript by the lowest subscript present (here, 1.66) gives:

$$C_{\frac{5.00}{1.66}}H_{\frac{13.3}{1.66}}O_{\frac{1.66}{1.66}} = C_{3.01}H_{8.01}O_{1.00} = C_3H_8O.$$

EMPIRICAL FORMULAS AND MOLECULAR FORMULAS:

Percentage composition data can only provide the empirical formula of a compound -- that is, the formula which expresses the ratios of the atoms present in the compound using the smallest whole numbers possible. To completely determine the molecular formula of a compound, the molecular weight must be determined and compared to the "empirical formula weight". Examples appear below:

Name of Compound	Empirical Formula	Empirical Formula Weight	Molecular Weight	Molecular Formula
Isopropyl Alcohol	C_3H_8O	60 g/mole	60 g/mole	C_3H_8O
Acetylene	CH	13 g/mole	26 g/mole	C_2H_2
Glucose	CH_2O	30 g/mole	180 g/mole	$C_6H_{12}O_6$

For isopropyl alcohol, the empirical formula weight is the same as the molecular weight, so the empirical formula is the same as the molecular formula. For acetylene, the two formula weights differ by a factor of two, so the molecular formula is just twice the empirical formula -- $(CH)_2 = C_2H_2$. For glucose, 180/30 = 6, so the molecular formula is $(CH_2O)_6 = C_6H_{12}O_6$.

Problem: Butane (the fluid in cigarette lighters), contains 82.66% C and 17.34% H by mass. The molecular weight of butane is approximately 58 g/mole. Calculate the molecular formula of butane.

Solution: 100.00 g of butane = 82.66 g of C and 17.34 g of H. Moles: 82.66/12.011 = 6.882 moles C, 17.34/1.0079 = 17.20 moles H. $C_{6.882}H_{17.20}$ can't be right, so dividing by 6.882 gives $C_{1.000}H_{2.500}$. Multiplying by 2 gives us whole numbers: C_2H_5 = empirical formula = 29 g/mole. Molecular formula = C_4H_{10}.

Alternative Solution: Multiply 58 g of butane by the percentages: 58 g x .8266 = 48 g of C, 58 g x .1734 = 10 g of H. Convert these into moles: $\frac{48 \text{ g of C}}{12 \text{ g/mole}}$ = 4 moles of C, $\frac{10 \text{ g of H}}{1 \text{ g/mole}}$ = 10 moles of H. Molecular formula = C_4H_{10}.

MOLECULES, IONS, AND POLYATOMIC IONS:

Some compounds are made of molecules, which are electrically neutral collections of two or more atoms that are joined in some way. These compounds are called covalent compounds, and are represented by molecular formulas. Water (H_2O) and sucrose ($C_{12}H_{22}O_{11}$, table sugar) are examples of covalent compounds. Some elements are also molecular in nature. For example, the oxygen in the air we breathe occurs naturally as molecules with the molecular formula O_2.

Other compounds are made of ions, which are electrically charged particles made from one or more atoms. These compounds are called ionic compounds, and are represented by empirical formulas. (There's no such thing as a "molecule" of an ionic compound, so it doesn't make much sense to refer to its formula as a "molecular" formula.) An example of an ionic compound is sodium chloride (NaCl, table salt), which is made up of two kinds of ions: Na^+ and Cl^-. Notice that the charge of the ion is written as a superscript. Ions with a positive charge are called cations; ions with a negative charge are called anions.

Polyatomic ions are ions which are made from more than one atom. Examples include the hydroxide ion, OH^-, which is found in lye (sodium hydroxide, NaOH), and the carbonate ion, CO_3^{2-}, which is found in limestone (calcium carbonate, $CaCO_3$). Some polyatomic ions contain three or more elements. An example is the acetate ion, $C_2H_3O_2^-$, which is found in vinegar (acetic acid, $HC_2H_3O_2$). Polyatomic ions may be either anions (like those above) or cations. An example of a polyatomic cation is the ammonium ion, NH_4^+, which is found in the fertilizer ammonium nitrate (NH_4NO_3, which also contains the nitrate ion, NO_3^-). Other common polyatomic ions include the sulfate ion (SO_4^{2-}), the phosphate ion (PO_4^{3-}), and the bicarbonate ion (HCO_3^-, which is found in baking soda (sodium bicarbonate, $NaHCO_3$)).

WRITING FORMULAS FOR IONIC COMPOUNDS:

The formulas of ionic compounds are written as empirical formulas -- that is, the smallest whole numbers are used which accurately describe the ratios of the ions present. For example, the formula for sodium chloride is written as NaCl, not Na_2Cl_2 or Na_3Cl_3. Simply "NaCl" is enough to show the one-to-one ratio of the sodium ions (Na^+) to the chloride ions (Cl^-).

Two things should be noted here. First, when chemists write the formulas of ionic compounds, they almost always write the symbol for the cation first and the symbol for the anion last. (Sodium chloride is "NaCl", not "ClNa". "ClNa" is not "wrong", but chemists prefer to write the symbol for the positive ion first to avoid confusion about which ion has which charge.) Second, the net electrical charge implied by the compound's formula is zero -- that is, the total of the positive charges is exactly cancelled by the total of the negative charges. This must be true for any ionic compound, since all compounds are electrically neutral. (In sodium chloride, the +1 charge of Na^+ is exactly cancelled by the -1 charge of Cl^-. This explains why the formula of sodium chloride is NaCl, not $NaCl_2$ or Na_2Cl. On the other hand, calcium chloride has the formula $CaCl_2$ due to the fact that calcium ions have a charge of +2. Therefore, two Cl^- ions are needed to exactly cancel one Ca^{2+} ion, and the formula is $CaCl_2$.)

When writing the formulas of compounds which contain polyatomic ions, it's a good idea to put parentheses around the formula of the polyatomic ion, especially if more than one polyatomic ion is present in the formula. For example, calcium carbonate is written as $CaCO_3$ (since only one CO_3^{2-} ion is present), but calcium hydroxide is written as $Ca(OH)_2$ (since two OH^- ions are present). Writing calcium hydroxide as CaO_2H_2 is not "wrong", but it's not as descriptive as $Ca(OH)_2$ -- the parentheses clearly show the presence of OH^- ions.

ACIDS, BASES, AND CHEMICAL REACTIONS:

Compounds which generate hydrogen ions (H^+) are called acids. Some well-known acids include hydrochloric acid (HCl, "stomach acid"), sulfuric acid (H_2SO_4, found in "acid rain"), and acetic acid ($HC_2H_3O_2$, vinegar). Notice that the formulas of these acids all begin with "H". This is usually a good "tip-off" that a compound might be an acid.

Compounds which generate hydroxide ions (OH^-) are called bases or alkalis. Some well-known bases include sodium hydroxide ($NaOH$, "lye") and magnesium hydroxide ($Mg(OH)_2$, "milk of magnesia"). Notice that the formulas of these bases both end with "OH". For ionic compounds, this is a good "tip-off" that a base is present.

Acids and bases tend to neutralize each other when they are mixed together -- that is, they tend to form new substances which are neither acidic nor basic. This is one example of a chemical reaction -- the changing of one or more substances into one or more new substances. Typically, an acid-base reaction consists of the H^+ ions from the acid neutralizing the OH^- ions from the base by combining with them to form molecules of water (HOH, or H_2O). For example, the reaction of hydrochloric acid with sodium hydroxide forms water and sodium chloride -- table salt. This reaction is summarized in the chemical equation:

$$HCl + NaOH \longrightarrow H_2O + NaCl$$

In this reaction, HCl and $NaOH$ are referred to as the reactants -- the substances which react with each other. The products (the new substances which are formed) are H_2O and $NaCl$. The arrow indicates the progress of the reaction, and can be read as "forms" or "produces" or "yields" or "gives". One way to "read" the above equation is to say that "hydrochloric acid and sodium hydroxide produce water and sodium chloride."

WRITING BALANCED EQUATIONS FOR CHEMICAL REACTIONS:

Suppose we wanted to write a chemical equation which describes the reaction between hydrogen gas and oxygen gas to form liquid water. One way to do this might be to simply write: $H + O \rightarrow H_2O$. However, this isn't the most best way to describe this reaction, since hydrogen and oxygen normally occur as <u>diatomic molecules</u> -- H_2 and O_2. Therefore, a somewhat better way to write this equation would be: $H_2 + O_2 \rightarrow H_2O$. But even this equation isn't perfect. Notice that there are <u>two</u> oxygen atoms among the <u>reactants</u>, but only <u>one</u> oxygen atom among the <u>products</u>. This seems to imply that an oxygen atom has "vanished into thin air", which is impossible according to Dalton's atomic theory.

The best way to write the above reaction is: $2\ H_2 + O_2 \rightarrow 2\ H_2O$. This equation states that <u>two</u> molecules of H_2 react with <u>one</u> molecule of O_2, producing <u>two</u> molecules of H_2O. (The "2" in front of the H_2 and the H_2O is called the <u>coefficient</u> of that formula. When the coefficient is "1", as it is for O_2 in this case, the "1" is usually omitted.) Note that in this equation, there are <u>four</u> hydrogen atoms among the reactants <u>and</u> the products, and <u>two</u> oxygen atoms are also found on <u>both</u> sides of the arrow. When the numbers of atoms of each element found on both sides of the arrow are the <u>same</u>, the equation is called <u>"balanced"</u>.

The usual way to balance an equation is to write the formulas first, then change the coefficients. (In the above case, we could have "balanced" the equation by changing a formula -- $H_2 + O_2 \rightarrow H_2O_2$ -- but this is the equation of a <u>different reaction</u>, not the formation of <u>water</u>!)

<u>Problem</u>: Balance the following equation: $Fe + Cl_2 \rightarrow FeCl_3$.

<u>Solution</u>: The reactants contain one less chlorine atom than the products do. Placing a "3" in front of the Cl_2 and a "2" in front of the $FeCl_3$ corrects this. Placing a "2" in front of the Fe gives: $2\ Fe + 3\ Cl_2 \rightarrow 2\ FeCl_3$.

COOKING, CHEMISTRY, AND STOICHIOMETRY:

Consider the following shortbread recipe[1]:

> "In a bowl, cream 1 cup of softened butter and 1/2 cup of sifted powdered sugar until light and fluffy. Stir in 2 cups of all-purpose flour and blend well. Gather up the dough and press it evenly over the bottom of a lightly-greased 9-inch square baking pan. Bake for 45 minutes at 325 °F. Transfer pan to wire rack and cut into 25 squares. Let cool in pan."

Since the above process describes chemical reactions -- after all, substances are being converted into new substances -- a chemist might record the information above using a chemical equation such as the one given below:

$$1\ \text{Bu} + 1/2\ \text{Su} + 2\ \text{Fl} \xrightarrow[45\ \text{min}]{163\ ^\circ\text{C}} 25\ \text{Sq}$$

where "Bu" = butter, "Su" = sugar, "Fl" = flour, and "Sq" = shortbread squares!

The equation above gives the same information as the recipe does about the reactions involved -- the relative amounts of the ingredients to use, the "reaction conditions" (temperature and baking time), and the number of shortbread squares produced -- but the recipe is obviously more useful in the kitchen!

The stoichiometry of a chemical reaction is simply the quantitative description of the relative amounts of the substances involved in the reaction. The difference for chemists is that "amounts" are measured in moles, not cups! Consider the following balanced equation: $PCl_5 + 4\ H_2O \rightarrow H_3PO_4 + 5\ HCl$. This equation tells us that one PCl_5 molecule and four water molecules will react to give one H_3PO_4 molecule and five HCl molecules. It also tells us that one mole of PCl_5 and four moles of water react to give one mole of H_3PO_4 and five moles of HCl. The balanced equation therefore allows us to think on both the small scale (molecules) and the large scale (moles) simultaneously. Of course, to actually do this reaction, a recipe (or "experimental procedure") would come in handy in lab!

---

1) "Easy Basics for Good Cooking", Sunset Books, Lane Publishing Co., Menlo Park, CA 94025.

SOLVING STOICHIOMETRY PROBLEMS:

The problem with chemical reactions is that they take place on the molecular level -- that is, only whole numbers of molecules react. Balanced chemical equations give us information about the numbers of molecules that react and the numbers of molecules that are formed -- that is, about the stoichiometry of the reaction. The problem is that in the laboratory, we can't measure out a particular number of molecules directly. All we can do is to measure out masses (in grams) of solids or liquids, or volumes (in liters) of liquids. Hence, we need to be able to convert these units into amounts in moles of molecules (and back again!) in order to be able to use the data that we obtain in the lab.

Typically, problems involving the stoichiometry of reactions are solved in the way outlined below. Consider the generic balanced equation below:

$$w\,A + x\,B \longrightarrow y\,C + z\,D$$

(This is to be read as "w moles of A and x moles of B react to form y moles of C and z moles of D". The equation must be balanced, since only the balanced equation represents the true ratios in which molecules react and form.)

Suppose you weigh out some mass of compound A and you want to find out what mass of compound B you need to weigh out to react completely with your sample of A. The first thing to do is to convert your mass of A (in grams) into a number of moles of A. This should be easy -- just divide by the molecular weight of A, in g/mole! Then, since A and B react in a mole ratio of w:x (look at the balanced equation!), multiply the amount of A (in moles) by x/w to obtain the amount of B (in moles) needed to react completely with your sample of A. (Here, the balanced equation is the conversion equation, and x/w is the conversion factor!) The final step is to convert this quantity back to a mass (in grams), which can be done by simply multiplying by the molecular weight of B (in g/mole).

A TYPICAL STOICHIOMETRY PROBLEM:

Problem: What mass of O_2 is needed to react completely with 56.8 g of NH_3 ? What masses of NO and H_2O are formed? The balanced equation is:

$$4\ NH_3 + 5\ O_2 \longrightarrow 4\ NO + 6\ H_2O$$

Solution: To find the mass of O_2 needed, first convert NH_3 to moles:

$$56.8\ g\ NH_3 \times \frac{1\ mole\ NH_3}{17.0304\ g\ NH_3} = 3.34\ moles\ NH_3.$$

Next, use the balanced equation to find the amount of O_2 needed:

$$3.34\ moles\ NH_3 \times \frac{5\ moles\ O_2}{4\ moles\ NH_3} = 4.17\ moles\ O_2.$$

Finally, convert this amount to a mass:

$$4.17\ moles\ O_2 \times \frac{31.9988\ g\ O_2}{1\ mole\ O_2} = 133\ g\ O_2.$$ (Note: 3 "sig figs"!)

The mass of NO formed is found in a similar fashion. The first step is the same as the first step above -- 3.34 moles of NH_3 are reacted. From there:

$$3.34\ moles\ NH_3 \times \frac{4\ moles\ NO}{4\ moles\ NH_3} = 3.34\ moles\ NO.$$

$$3.34\ moles\ NO \times \frac{30.0061\ g\ NO}{1\ mole\ NO} = 100\ g\ NO.$$ (Again, 3 "sig figs"!)

The mass of H_2O formed is found similarly. This time, only the conversion factors are shown, but the stepwise logic is the same!

$$56.8\ g\ NH_3 \times \frac{1\ mole\ NH_3}{17.0304\ g\ NH_3} \times \frac{6\ moles\ H_2O}{4\ moles\ NH_3} \times \frac{18.0152\ g\ H_2O}{1\ mole\ H_2O} = 90.1\ g\ H_2O.$$

To check your work, apply the Law of Conservation of Mass. The total mass of the products should be the same as the total mass of the reactants!

Reactants	Products
56.8 g NH_3	100 g NO
+ 133 g O_2	+ 90.1 g H_2O
190 g reactants	190 g products

(The masses are equal, given that each mass only contains three significant figures!)

LIMITING REACTANTS:

Problem: How much $FeCl_3$ can be formed from the reaction of 10.0 g of Fe and 20.0 g of Cl_2 ? The balanced equation is: $2\ Fe + 3\ Cl_2 \rightarrow 2\ FeCl_3$.

Solution: 29.0 grams! Not 30.0 grams (10.0 g + 20.0 g). Why?

Consider this as a stoichiometry problem: What mass of Cl_2 is needed to react with 10.0 g of iron? The calculations are shown below.

$$10.0\ g\ Fe \times \frac{1\ mole\ Fe}{55.847\ g\ Fe} \times \frac{3\ moles\ Cl_2}{2\ moles\ Fe} \times \frac{70.906\ g\ Cl_2}{1\ mole\ Cl_2} = 19.0\ g\ Cl_2.$$

What happens in the reaction is that the 10.0 g of iron reacts with the 19.0 g of chlorine calculated above, giving 29.0 g of $FeCl_3$ and leaving 1.0 g of chlorine unreacted, since there is no more iron for it to react with. In this case, iron is referred to as the limiting reactant (or limiting reagent), since it is used up before all of the chlorine can react and thereby sets an upper limit on the amount of $FeCl_3$ that can be formed.

Problem: For the reaction above, if 10.0 grams of iron and 10.0 grams of chlorine are allowed to react, how much $FeCl_3$ can be formed?

Solution: The above calculation shows that 10.0 g of Fe requires 19.0 g of Cl_2 to react completely. Since only 10.0 g of Cl_2 are available here, the Cl_2 will be consumed first. Therefore, Cl_2 is the limiting reagent. The $FeCl_3$ calculation must be based on the amount of the limiting reagent present. Hence:

$$10.0\ g\ Cl_2 \times \frac{1\ mole\ Cl_2}{70.906\ g\ Cl_2} \times \frac{2\ moles\ FeCl_3}{3\ moles\ Cl_2} \times \frac{162.206\ g\ FeCl_3}{1\ mole\ FeCl_3} = 15.3\ g\ FeCl_3.$$

Problem: For the reaction above, if 10.0 g of iron and 15.0 g of chlorine are allowed to react, which is the limiting reagent?

Solution: Don't be fooled by the masses! 15.0 g is less than the 19.0 g calculated above, so Cl_2 is the limiting reagent in this case.

THEORETICAL YIELD AND PERCENTAGE YIELD:

The theoretical yield of a reaction refers to the amount of product that can be formed if the limiting reactant is completely converted into the product. Theoretical yields are expressed in either grams or moles.

Problem: Calculate the theoretical yield of $FeCl_3$ if 10.0 grams of Fe and 15.0 grams of Cl_2 are allowed to react. The balanced equation is:

$$2\ Fe + 3\ Cl_2 \rightarrow 2\ FeCl_3$$

Solution: Which is the limiting reactant? In this case, it's Cl_2:

$$15.0\ g\ Cl_2 \times \frac{1\ mole\ Cl_2}{70.906\ g\ Cl_2} \times \frac{2\ moles\ Fe}{3\ moles\ Cl_2} \times \frac{55.847\ g\ Fe}{1\ mole\ Fe} = 7.88\ g\ Fe.$$

This is the amount of iron needed to react with 15.0 g of chlorine. 10.0 g of iron is available. Iron is present in excess, so chlorine is limiting.

Now, calculate the theoretical yield, using the limiting reagent:

$$15.0\ g\ Cl_2 \times \frac{1\ mole\ Cl_2}{70.906\ g\ Cl_2} \times \frac{2\ moles\ FeCl_3}{3\ moles\ Cl_2} \times \frac{162.206\ g\ FeCl_3}{1\ mole\ FeCl_3} = 22.9\ g\ FeCl_3.$$

The yield of a reaction is just the amount of product that is actually obtained from a reaction. The yield is determined by experiment; the theoretical yield is found by calculation. The percentage yield of a reaction is simply the ratio of the yield of a reaction to its theoretical yield, multiplied by 100%.

Problem: A student obtained a yield of 20.0 g of $FeCl_3$ when she reacted 10.0 g of Fe with 15.0 g of Cl_2. Calculate her percentage yield.

Solution: $$\text{Percentage Yield} = \frac{\text{Yield} \times 100\%}{\text{Theoretical Yield}} = \frac{20.0\ g}{22.9\ g} \times 100\%$$

$$= 87.3\%\ \text{yield}.$$

Percentage yields of less than 100% are common. Side reactions, which give products other than the desired product, usually account for the "missing" mass. (A possible side reaction in this case is: $Fe + Cl_2 \rightarrow FeCl_2$.)

SOME TERMS USED TO DESCRIBE SOLUTIONS:

Solutions are homogeneous mixtures of two or more substances. Coffee, air, and brass are all examples of solutions. The substance which is present in the largest amount in a solution is usually called the solvent. Water is a common solvent (sometimes called the "universal solvent"), and solutions in which water is the solvent are called aqueous solutions (from the Latin word "aqua", meaning "water"). The substances which are dissolved in the solvent are called solutes. In soft drinks, the solutes include sugar, caffeine, and carbon dioxide.

The concentration of a solution is a description of the ratio of the amounts of solutes and solvents present. Qualitatively, solutions may be either dilute (solute/solvent ratio is relatively small) or concentrated (solute/solvent ratio is large). A solution which is so concentrated that no more solute will dissolve in it is called a saturated solution. However, it is sometimes possible to make a solution even more concentrated than this by heating the saturated solution (heating a solution usually increases the tendency of the solute to dissolve), adding more solute, stirring the solution to dissolve the solute, and carefully cooling the solution back to the original temperature. Solutions such as these are called supersaturated solutions.

Concentrations of solutions are usually expressed in quantitative terms for laboratory purposes. The most common of these expressions is the molarity of a solution, which is defined as the amount of solute present per liter of solution. "Amount of solute" is measured in moles, so concentrations expressed in this way are useful in the laboratory, where calculations involving moles are common. Molarity is abbreviated by "M". For example, a saturated solution of calcium hydroxide has a concentration of 0.0250 M, since 0.0250 moles of $Ca(OH)_2$ will dissolve in enough water to make 1.00 liter of solution at room temperature.

CALCULATIONS INVOLVING CONCENTRATIONS:

<u>Problem</u>: How would you make 250.0 mL of a 0.100 M solution of $AgNO_3$?

<u>Solution</u>: Since "M" means "moles per liter", multiply the volume of the solution (in liters) by the concentration to obtain the amount (in moles) of $AgNO_3$ needed. Converting this to a mass in grams, we obtain the following:

$$250.0 \text{ mL} \times \frac{1 \text{ L}}{1000 \text{ mL}} \times \frac{0.100 \text{ moles } AgNO_3}{1 \text{ L of solution}} \times \frac{169.873 \text{ g } AgNO_3}{1 \text{ mole } AgNO_3} = 4.25 \text{ g of } AgNO_3.$$

To actually make this solution, 4.25 g of $AgNO_3$ would be weighed out and placed in a 250-mL <u>volumetric</u> flask. (A volumetric flask is a vessel which has been calibrated to show the <u>exact</u> volume of liquid present in the flask.) Distilled water would then be added to the flask until the $AgNO_3$ dissolves and the total volume of the solution is exactly 250.0 mL. (This is <u>not</u> the same as adding 250.0 mL of water to the $AgNO_3$. Actually, <u>less</u> water would be required.)

<u>Problem</u>: How would you make 100.0 mL of a 2.90 M solution of HCl? (Solutions of HCl with a concentration of 11.6 M are commercially available.)

<u>Solution</u>: Determine the amount (in moles) of HCl present in the desired solution. Then, calculate the volume of the concentrated solution needed to provide this amount. The difference in volume will be made up by adding distilled water to make the concentrated solution more dilute. Hence:

$$100.0 \text{ mL} \times \frac{1 \text{ L}}{1000 \text{ mL}} \times \frac{2.90 \text{ moles HCl}}{1 \text{ L solution}} \times \frac{1 \text{ L solution}}{11.6 \text{ moles HCl}} \times \frac{1000 \text{ mL}}{1 \text{ L}}$$

$= 25.0$ mL of concentrated (11.6 M) HCl needed.

To make this solution, add about 50 mL of distilled water to a 100-mL volumetric flask, then add 25.0 mL of 11.6 M HCl. (Acid should be poured into water, not the other way around -- if anything splashes, let it be water!) More distilled water should then be added to the volumetric flask to make the final volume exactly equal to 100.0 mL.

STOICHIOMETRY OF REACTIONS IN SOLUTION:

Problem: A scientist knocks over a volumetric flask containing 100.0 mL of a 2.90 M solution of HCl, spilling the solution on the lab bench. Since HCl is an acid, a base must be used to neutralize it, but the only base available is a 1.97 M solution of Na_2CO_3. What volume (in mL) of the Na_2CO_3 solution is needed to neutralize the HCl solution? The balanced equation of the reaction is:

$$2\ HCl_{(aq)} + Na_2CO_{3\ (aq)} \rightarrow 2\ NaCl_{(aq)} + H_2O_{(l)} + CO_{2\ (g)}$$

where the abbreviations "aq", "l", and "g" stand for "aqueous solution", "liquid", and "gas", respectively. (Solids are indicated by the subscript "s".)

Solution: This problem is solved in the same way any stoichiometry problem must be solved -- namely, using moles. First, convert the known volume and concentration of HCl into an amount in moles:

$$100.0\ mL \times \frac{1\ L}{1000\ mL} \times \frac{2.90\ moles\ HCl}{1\ L\ solution} = 0.290\ moles\ of\ HCl.$$

Next, use the mole ratio of HCl to Na_2CO_3 to calculate the amount of Na_2CO_3 (in moles) needed to neutralize the above amount of HCl:

$$0.290\ moles\ of\ HCl \times \frac{1\ mole\ of\ Na_2CO_3}{2\ moles\ of\ HCl} = 0.145\ moles\ of\ Na_2CO_3\ needed.$$

Finally, use the concentration of Na_2CO_3 to convert this into the desired volume, expressed in mL:

$$0.145\ moles\ Na_2CO_3 \times \frac{1\ L\ of\ solution}{1.97\ moles\ Na_2CO_3} \times \frac{1000\ mL}{1\ L} = 73.6\ mL\ of\ Na_2CO_3.$$

(Of course, had this been an actual emergency, the scientist should neutralize the acid spill first and do calculations later!)

Notice that in the above sequence of calculations, we divided by 1000 and multiplied by 1000. Since these two steps cancel each other out, they can be safely omitted by those who wish to work in millimoles. (mL x M = mmol!)

VOLUMETRIC ANALYSIS USING TITRATIONS:

Volumetric analysis refers to any kind of analytical method based on measuring volumes. A typical laboratory example is a titration, in which two solutions are combined until the reactants in each solution are exactly consumed.

The most common type of titration involves the neutralization of an acid by a base. Usually, a solution of a known base is made and standardized by titrating a known volume of the base with a known mass or volume of a known acid. The stoichiometry of the acid-base reaction is then used to calculate the concentration of the basic solution, which is then used to titrate an unknown acid. The results of the titration provide information about the unknown acid.

An indicator is used to determine the point at which the reactants in each solution are exactly consumed. (This point is called the end point, since the titration ends when this point is reached.) Indicators are substances which change color when some property of the solution in which they are dissolved changes. Typical indicators for acid-base titrations include phenolphthalein (which is colorless in acidic solutions but red in basic solutions) and bromothymol blue (which is yellow in acids but blue in bases). The appearance of an intermediate color (pink for phenolphthalein or green for bromothymol blue) signals the end point, at which the acid and base have been neutralized.

Problem: A 229-mg sample of an unknown acid was titrated using 29.70 mL of 0.0965 M NaOH solution. What is the molecular weight of the acid? The balanced equation of the reaction which occurs is: $HX + NaOH \rightarrow NaX + H_2O$.

Solution: The necessary conversion factors are shown below.

$$29.70 \text{ mL NaOH} \times \frac{1 \text{ L}}{1000 \text{ mL}} \times \frac{0.0965 \text{ mol NaOH}}{1 \text{ L of solution}} \times \frac{1 \text{ mol HX}}{1 \text{ mol NaOH}} = 2.87 \times 10^{-3} \text{ moles of HX.}$$

$$\text{Molecular Weight} = \frac{229 \text{ mg of HX}}{2.87 \times 10^{-3} \text{ mol HX}} \times \frac{1 \text{ g}}{1000 \text{ mg}} = 79.9 \text{ g/mole.}$$

ENERGY AND ITS MEASUREMENT:

When you lift a heavy object, you oppose the force of gravity which pulls that object toward the center of the earth. This is one example of doing work. Work is done whenever a force is exerted against an opposing force. You are able to do work because you have energy. Energy is difficult to define, but you can think of it as something that an object has if it is able to do work.

Energy can have many forms, some of which should be familiar. For example, kinetic energy (sometimes called mechanical energy) is energy of motion, and is defined by the equation $KE = mv^2/2$, where m is an object's mass and v is its velocity. Potential energy is "stored" energy. For example, chemical energy is a form of potential energy which is stored within compounds in the form of attractive forces (called chemical bonds) which connect the atoms or ions of a compound to each other.

All of the different forms of energy can be totally converted into heat. Heat is a familiar form of energy -- you learned what "hot" and "cold" meant when you were a child! Heat is that form of energy which flows from "hot" objects toward "cold" objects. Heat is not the same as temperature -- temperature is a measurement of the intensity of the heat present in an object. Besides, the two are measured using different units: temperature is measured in degrees, but energy is measured in joules or calories. Since $KE = mv^2/2$, the units of energy must have the dimensions of mass x velocity^2. One joule = 1 kg m^2/sec^2. (An object with a mass of 2 kg moving at a velocity of 1 m/sec has a kinetic energy of 1 J.) One calorie = 4.184 joules. (This is a defined number, containing an infinite number of "sig figs"!) One calorie is also the amount of heat needed to raise the temperature of one gram of water by one Celsius degree. (The "calories" on food labels are actually kilocalories -- be aware of this!)

THERMOCHEMISTRY:

The potential energy of a system increases whenever objects which repel each other are brought closer together, or whenever objects which attract each other are separated. Compounds exist because the atoms or ions from which they are made attract each other. These attractive forces between atoms or ions in compounds are a form of potential energy called chemical bonds. Dalton's atomic theory states that atoms or ions are rearranged during a chemical reaction. As an example, consider the "thermite" reaction, which has been used for welding:

$$2\ Al + Fe_2O_3 \longrightarrow 2\ Fe + Al_2O_3$$

Notice that in the course of this reaction, the oxygen atoms become separated from the iron atoms and end up attached to the aluminum atoms. One way to think of this as occurring is in a two-step process, as shown below:

$$\text{Step One: } 2\ Al + Fe_2O_3 \longrightarrow 2\ Al + 2\ Fe + 3\ O$$

$$\text{Step Two: } 2\ Al + 2\ Fe + 3\ O \longrightarrow 2\ Fe + Al_2O_3$$

In the first step, the attractive forces between the Fe and O atoms are overcome. As these atoms are separated, the chemical bonds are "broken", and the potential energy of the system increases. In Step Two, new chemical bonds are formed due to the attraction of the Al atoms for the O atoms. As these atoms are brought closer together, the potential energy of the system decreases.

Changes in potential energy are usually observed as the consumption or release of heat. Therefore, the study of the changes in energy which occur during chemical reactions is called thermochemistry (from the Greek "therme", meaning "heat"). In Step One, the reaction consumes energy. Energy-consuming processes are called endothermic (from the Greek "endon", meaning "within"). In Step Two, the reaction releases energy. Energy-releasing processes are called exothermic (from the Greek "exo", meaning "out").

ENTHALPY OF REACTIONS:

The amount of heat released or consumed in a chemical reaction is called the heat of reaction, and is given the symbol q_r, where "q" means "heat" and "r" means "reaction". The value of q_r is a positive number for an endothermic (energy-consuming) process, and is a negative number for an exothermic (energy-releasing) process. Heats of reaction are measured in either joules or calories.

A chemical reaction which is carried out in a vessel that is open to the air is exposed to atmospheric pressure, which usually remains roughly constant over short periods of time. The potential energy of a system at constant pressure is called the enthalpy of the system. When a chemical reaction occurs, the potential energy (chemical energy) of the system changes, so the enthalpy changes. This change in enthalpy is simply the heat of reaction. In equation form, this is written: $\Delta H_r = q_r$, where "Δ" (the Greek letter delta) means "change" and "H" means "enthalpy".

We can tell whether a chemical reaction is exothermic or endothermic by determining whether it feels "hot" (gives off heat) or "cold" (consumes heat). That is, we observe changes in temperature. To measure temperature changes, all we need to know is the initial temperature and the final temperature of the reaction mixture, and then simply subtract the two: $\Delta T = T_{final} - T_{initial}$. Temperature is one of several quantities which are called state functions, since changes in temperature depend only on the initial and final states of the system being studied. Enthalpy is also a state function -- the enthalpy change of a reaction is simply the difference in enthalpy between the products (final state) and the reactants (initial state): $\Delta H_r = H_{products} - H_{reactants} = q_r$. This is true regardless of the actual route by which the reaction gets from the reactants to the products -- all that matters are the initial and final states.

STANDARD HEATS OF REACTIONS:

The amount of heat consumed or released by a given chemical reaction varies with the pressure and temperature at which the reaction occurs. Scientists who measure heats of reactions have chosen a set of standard conditions for measuring heats of reactions. These standard conditions are a temperature of 25.0 ^{o}C (77.0 ^{o}F, approximately "room temperature") and an atmospheric pressure of 1.0 atmosphere (29.92 inches of mercury, which are the units used to describe atmospheric pressure in weather reports).

The standard heat of reaction is simply the heat of reaction measured under the above standard conditions. A thermochemical equation is an equation of a reaction which also includes the standard heat of reaction. For example:

$$2\ Fe_{(s)} + \frac{3}{2} O_{2\ (g)} \longrightarrow Fe_2O_{3\ (s)} \qquad \Delta H^{o}_{r} = -196.5\ kcal$$ [1]

(The superscript "o" in the symbol "ΔH^{o}_{r}" indicates that this is a standard heat of reaction, measured at 25 ^{o}C and a pressure of 1.0 atmosphere.)

Notice that under the standard conditions, iron is a solid and oxygen is a gas. These are the most stable physical forms of iron and oxygen at 25.0 ^{o}C and 1.0 atmosphere. These elements are therefore said to be in their standard states -- that is, the physical states in which they would normally occur. The above reaction therefore represents the formation of solid Fe_2O_3 from the elements of which it is composed, in their standard states. The standard heat of formation of a compound is simply the heat of reaction for the reaction where the compound is formed from its elements in their standard states. The symbol "ΔH^{o}_{f}", where "f" means "formation", is used to indicate the standard heat of formation. Hence, the information in the above thermochemical equation can be summarized by saying that for $Fe_2O_{3\ (s)}$, $\Delta H^{o}_{f} = -196.5$ kcal.[1]

1) CRC Handbook of Chemistry and Physics, 56th Edition, 1975-1976.

HESS' LAW:

Consider the thermite reaction, which has been used for welding:

$$Fe_2O_3\,(s) + 2\,Al\,(s) \rightarrow Al_2O_3\,(s) + 2\,Fe\,(s) \quad \Delta H_r^o = -202.6 \text{ kcal}$$

Since enthalpy is a state function, the above value of H depends on just the initial and final states of the reaction, and not on the actual pathway by which the reaction gets from the reactants to the products. Therefore, we can mentally picture this reaction occurring by any pathway we choose. One possible reaction pathway is the decomposition of the reactants into the elements from which they are made, followed by the reaction of those elements to form products:

Step One: $Fe_2O_3\,(s) \rightarrow 2\,Fe\,(s) + \frac{3}{2}\,O_2\,(g) \quad \Delta H_r^o = +196.5 \text{ kcal}$[1]

Step Two: $2\,Al\,(s) + \frac{3}{2}\,O_2\,(g) \rightarrow Al_2O_3\,(s) \quad \Delta H_r^o = -399.09 \text{ kcal}$[1]

Note that -202.6 kcal = (+196.5 kcal) + (-399.09 kcal). Hess' Law states that for any chemical reaction that can be written as if it occurs in a stepwise fashion, the enthalpy change of the reaction equals the sum of the enthalpy changes of the individual steps. In equation form, $\Delta H_r = \Delta H_{Step\ 1} + \Delta H_{Step\ 2} + \Delta H_{Step\ 3} + \ldots$

Step Two above is simply the equation of the formation of Al_2O_3 from its elements in their standard states. Therefore, ΔH_r^o for this reaction is just the standard heat of formation (ΔH_f^o) of Al_2O_3. Step One is similar, except the equation is reversed. ΔH_f^o for Fe_2O_3 is -196.5 kcal. From this we can see that reversing the reaction simply changes the sign (+ or -) of the enthalpy change. Therefore, if the standard heats of formation of the reactants and products of a chemical reaction are available, then heats of reactions can be found using Hess' Law in its alternate form: $\Delta H_r^o = \Delta H_{f\,(products)}^o - \Delta H_{f\,(reactants)}^o$. This form is often useful, since values of ΔH_f^o have been compiled in reference texts.[1]

---

1) CRC Handbook of Chemistry and Physics, 56th Edition, 1975-1976.

CALCULATING STANDARD HEATS OF REACTIONS USING HESS' LAW:

The standard heats of formation of many compounds have been measured and recorded in reference texts.[1] Therefore, it is often possible to calculate the standard heat of reaction without doing the reaction. Examples appear below.

Reaction: $HCl_{(g)} + NaOH_{(s)} \rightarrow NaCl_{(s)} + H_2O_{(l)}$

To calculate ΔH_r^o for this reaction, just look up the standard heats of formation of the reactants and products in reference tables, then apply Hess' Law in the form: $\Delta H_r^o = \Delta H_f^o{}_{(products)} - \Delta H_f^o{}_{(reactants)}$. ΔH_f^o values[1]:

$HCl_{(g)}$ = -22.06 kcal/mole. $NaCl_{(s)}$ = -98.23 kcal/mole.

$NaOH_{(s)}$ = -101.99 kcal/mole. $H_2O_{(l)}$ = -68.32 kcal/mole.

(Note: Standard heats of formation are usually given as "kcal/mole" because the assumption is that one mole of the compound is being formed from its elements.)

$$\Delta H_r^o = \Delta H_f^o{}_{(products)} - \Delta H_f^o{}_{(reactants)}$$
$$= (-68.32 \text{ kcal/mole}) + (-98.23 \text{ kcal/mole})$$
$$- [(-22.06 \text{ kcal/mole}) + (-101.99 \text{ kcal/mole})] = -42.50 \frac{\text{kcal}}{\text{mole}}.$$

Reaction: $CH_{4\,(g)} + 2\,O_{2\,(g)} \rightarrow CO_{2\,(g)} + 2\,H_2O_{(g)}$

The calculation is similar to the one above, except we now have some coefficients which are not equal to one. ΔH_f^o values must be multiplied by the coefficients in this case. Note also that the ΔH_f^o values for $H_2O_{(l)}$ and $H_2O_{(g)}$ are different. Also, the value of ΔH_f^o for $O_{2\,(g)}$ is zero. This makes sense -- oxygen is an element, and since zero heat is required to convert an element into itself, the standard heat of formation of any element is zero. ΔH_f^o values[1]:

$CH_{4\,(g)}$ = -17.889 kcal/mole. $CO_{2\,(g)}$ = -94.052 kcal/mole.

$O_{2\,(g)}$ = 0.00 kcal/mole. $H_2O_{(g)}$ = -57.80 kcal/mole.

$$\Delta H_r^o = (-94.052) + (2)(-57.80) - [(2)(0) + (-17.889)] = -191.76 \text{ kcal}.$$

---

1) CRC Handbook of Chemistry and Physics, 56th Edition, 1975-1976.

CALORIMETRY:

Changes in energy cannot be measured directly in the laboratory. What can be measured directly are the changes in temperature caused by exothermic or endothermic reactions. The science of measuring these temperature changes and using them to calculate the heats of chemical reactions (or the amounts of heat released or consumed by other processes) is called calorimetry.

All calorimetry experiments are based on the Law of Conservation of Energy, which states that energy is never created or destroyed, but simply changed from one form into another or transferred from one place to another. Simply put, if one object loses heat, another object must gain heat. Since the Law of Conservation of Energy (sometimes called the First Law of Thermodynamics) states that the total amount of energy present in the universe remains constant, the amount of heat lost by an object must equal the amount of heat gained by other objects. In equation form, this is written as: $q_{lost} = -q_{gained}$, where "q" means "heat" and the "-" is present because loss of heat is an exothermic process, for which q is negative, while gain of heat is an endothermic process, for which q is positive. The "-" sign just makes the other signs come out the way they should.

For precise experimental work, scientists often use an instrument called a "bomb" calorimeter. (It doesn't explode!) The "bomb" is a stainless steel vessel in which the reaction occurs. The "bomb" is usually sealed and immersed in an insulated tub of water, whose temperature changes are then noted. Heats of combustion are usually measured in a "bomb" calorimeter. For scientists on a limited budget, a styrofoam coffee cup provides reasonably good insulation against heat loss. "Coffee-cup" calorimeter experiments work well for reactions that take place in solution, giving reasonably good results. Precisely measuring temperature is important to the success or failure of a calorimetry experiment.

HEAT CAPACITY AND SPECIFIC HEAT:

We can convert temperature changes into amounts of energy if we know the heat capacity of the object whose temperature is changing. The heat capacity of an object is simply the amount of heat needed to raise the object's temperature by 1.00 ^{o}C. Heat capacity is measured in units of energy per degree, such as J/K or cal/^{o}C. Heat capacity is an extensive property of a substance, but intensive properties are generally more useful, so the "heat capacity per gram" of a substance is used frequently. This property is called the specific heat of a substance, and it is defined as the amount of heat needed to raise the temperature of 1.00 gram of a substance by 1.00 ^{o}C. Specific heat is measured in units of energy per gram per degree, such as $\frac{cal}{g\ K}$ or $\frac{J}{g\ ^{o}C}$.

Problem: A 70.0-gram piece of metal was heated to 77.0 ^{o}C and placed in 100.0 grams of water at 22.0 ^{o}C in a "coffee-cup" calorimeter. The metal and the water came to the same temperature at 27.0 ^{o}C. Calculate the heat capacity and the specific heat of the metal.

Solution: The first step is to figure out the amount of heat that was transferred from the metal to the water. This is easily done if you remember that the specific heat of water is 1.00 $\frac{cal}{g\ ^{o}C}$. (One calorie raises the temperature of one gram of water by one Celsius degree.) Since we know the mass of the water (100.0 g) and its temperature change ($\Delta T = T_f - T_i$ = 27.0 ^{o}C - 22.0 ^{o}C = 5.0 ^{o}C), we can calculate the amount of heat transferred by cancelling the proper units:

$$q_{water} = 1.00\ \frac{cal}{g\ ^{o}C} \times 100.0\ g \times 5.0\ ^{o}C = 500\ \text{calories.}$$ (Two sig figs.)

Now, knowing that $q_{metal} = -q_{water}$ = -500 calories, and $\Delta T_{metal} = T_f - T_i$ = 27.0 ^{o}C - 77.0 ^{o}C = -50.0 ^{o}C, we can calculate the above quantities as shown:

Heat Capacity = $\frac{-500\ cal}{-50.0\ ^{o}C}$ = 10 cal/^{o}C. (2 sig figs)

Specific Heat = $\frac{-500\ calories}{(70.0\ g)(-50.0\ ^{o}C)}$ = 0.14 cal/g ^{o}C. (2 sig figs)

DETERMINING ENTHALPY CHANGES BY CALORIMETRY:

<u>Problem</u>: A sample of silver nitrate ($AgNO_3$) weighing 8.50 g was dissolved in 91.5 g of water in a "coffee-cup" calorimeter. The temperature of the water was 23.4 ^oC originally, but the addition of the $AgNO_3$ produced a drop in temperature of the solution, reaching a minimum temperature of 20.7 ^oC. Calculate the <u>molar heat of solution</u> of $AgNO_3$ -- that is, ΔH_r for the reaction: $AgNO_{3\,(s)} \longrightarrow AgNO_{3\,(aq)}$. Express your answer in kilocalories per mole of $AgNO_3$. Assume that the heat capacity of the calorimeter is zero and that the specific heat of the $AgNO_3$ solution is 1.00 $\frac{cal}{g\ ^oC}$.

<u>Solution</u>: In <u>any</u> calorimetry experiment, $q_{lost} = -q_{gained}$. In <u>this</u> experiment, the solution is <u>losing</u> heat (as seen by the drop in temperature), so the reaction must be <u>gaining</u> heat -- it is an <u>endothermic</u> reaction. Hence, the above equation can be rewritten: $q_{reaction} = -q_{solution}$. To determine the amount of heat transferred ($q_{solution}$), use the information supplied above:

Mass of Solution = 100.0 grams (8.50 g $AgNO_3$ + 91.5 g water)

$\Delta T_{solution} = T_f - T_i = 20.7\ ^oC - 23.4\ ^oC = -2.7\ ^oC$

$q_{solution} = 1.00\ \frac{cal}{g\ ^oC} \times 100.0\ g \times (-2.7\ ^oC) = -270\ cal.$ (2 sig figs.)

Therefore, $q_{reaction} = -q_{solution} = +270$ calories. To express this in kcal/mole, we need to calculate the amount of $AgNO_3$ (in moles) that was used.

$$8.50\ g\ AgNO_3 \times \frac{1\ mole\ AgNO_3}{169.873\ g\ AgNO_3} = 0.0500\ moles\ AgNO_3$$

$$\Delta H_r^o = \frac{+270\ calories}{0.0500\ moles\ AgNO_3} \times \frac{1\ kcal}{1000\ cal} = +5.4\ kcal/mole\ of\ AgNO_3.$$ [1]

Note that the result is a <u>positive</u> number, which it should be for an <u>endothermic</u> process. The assumption that the heat capacity of the calorimeter is zero was made to simplify the calculations -- there was no "$q_{calorimeter}$" term present.

1) CRC Handbook of Chemistry and Physics, 56th Edition, 1975-1976.

CALORIMETRY EXPERIMENTS IN FOOD CHEMISTRY:

Problem: A teaspoonful of sugar (about 5.0 grams) was placed in a "bomb" calorimeter which had a heat capacity of 1.00 kJ/^{o}C. The calorimeter was immersed in a water bath containing 1.00 L of water. The sugar was combusted completely in the "bomb", causing the temperature of the water to rise from its initial value of 24.96 ^{o}C to a maximum value of 40.96 ^{o}C. Calculate the amount of energy released by this reaction, and compare your result to the advertiser's claim that sugar contains "only sixteen calories per teaspoon".

Solution: As with any calorimetry problem, $q_{lost} = -q_{gained}$. In this case, the reaction releases heat, which is absorbed by the calorimeter and the water. Therefore, we can rewrite the above equation in the following way:

$$q_{reaction} = -[q_{water} + q_{calorimeter}]$$

Now, we can solve for the values of q_{water} and $q_{calorimeter}$ using the data above:

$$\Delta T_{water} = \Delta T_{calorimeter} = T_f - T_i = 40.96\ ^{o}C - 24.96\ ^{o}C = +16.00\ ^{o}C.$$

$$q_{calorimeter} = (1.00\ kJ/^{o}C) \times (+16.00\ ^{o}C) \times \frac{1.00\ kcal}{4.184\ kJ} = +3.82\ kcal.$$

$$q_{water} = 1.00\ \frac{calorie}{gram\ ^{o}C} \times (+16.00\ ^{o}C) \times 1.00\ L \times 1000\ mL/L \times 1.00\ g/mL$$

$$= 16{,}000\ cal = 16.0\ kcal.$$

Finally, using the equation above, we can solve for $q_{reaction}$:

$$q_{reaction} = -[q_{water} + q_{calorimeter}] = -(16.0\ kcal + 3.82\ kcal)$$

$$= -19.8\ kcal.\quad \text{(Three sig figs.)}$$

This is a little bit higher than the advertiser's claim of 16 "Calories" (food "Calories" are actually kilocalories) per teaspoon (level or heaping teaspoon?). This also works out to about 4.0 "Calories" per gram of sugar. This is a typical "Calorie content" (heat of combustion!) for a carbohydrate, which sugar is. (Sugar = $C_{12}H_{22}O_{11}$ = $C_{12}(H_2O)_{11}$ = "carbon-hydrate"!) Proteins also contain about 4.0 calories per gram, while fats contain over 9.0 calories per gram.

THE ORIGINS OF THE PERIODIC TABLE AND THE PERIODIC LAW:

Elements such as iron and sulfur have been known for millenia, but only in the last few centuries have scientists applied themselves to the study of the properties of the elements. By the middle of the 19th century, many of the more common elements had been studied and characterized, but this information was just a haphazard collection of observations -- it had not been organized in any particular way. In the 1860's, the Russian chemist Dmitri Mendeleev, while in the process of writing a chemistry textbook, wrote down the known information about the elements on index cards. He then noticed that if he arranged the cards in order of increasing atomic weight of the elements, the properties of the elements varied in a periodic way -- that is, every so often an element's properties would be similar to those of another element which had been previously encountered.

Mendeleev then arranged his index cards in a rectangular array, so that elements with similar properties were placed in the same column, and elements with different properties were placed in different columns. This arrangement left him with some "holes" in his periodic table, which he suspected corresponded to elements which had yet to be discovered. This hypothesis was shown to be correct; for example, the "hole" between silicon and tin was filled by germanium in 1886.

In 1912, the British physicist Henry Moseley discovered a method by which the atomic numbers (the numbers of protons or electrons present in atoms) of the elements could be accurately determined. Thereafter, it was found that a periodic table in which the elements were arranged in horizontal rows according to increasing atomic number was more consistent than Mendeleev's 1869 table. The modern version of the periodic law states that the chemical and physical properties of the elements vary in a periodic way as their atomic numbers increase. (Moseley died in battle during World War I at the age of 27.)

THE MODERN PERIODIC TABLE:

The Periodic Table of the Elements as we know it today is not quite the same as Mendeleev's periodic table. Today's Periodic Table consists of seven horizontal rows (sometimes called periods) numbered 1 to 7 from top to bottom. Each row contains between two and thirty-two elements. (The two rows which often appear at the bottom of the Periodic Table are actually part of Rows 6 and 7. The first of these rows, elements 58 through 71, are called the lanthanides after the 57th element, lanthanum. The other row, elements 90 through 103, are called the actinides after the 89th element, actinium.)

The elements are also arranged in vertical columns which are often called groups. Each group is designated by a Roman numeral and a letter. The groups which contain elements 1 through 20 are called the main groups and are assigned the letter "A". The groups which contain elements 21 through 30 are called the transition metals and are assigned the letter "B". The "A" groups are assigned the Roman numerals I through VIII going from left to right, but the "B" groups start with III on the left, increase through VIII (which describes the three-group block which contains elements 26-28, 44-46, and 76-78), and finishes with I and II. Thus, magnesium (Mg) is in Group IIA, chlorine (Cl) is in Group VIIA, chromium (Cr) is in Group VIB, and copper (Cu) is in Group IB.[1]

Some of the groups are referred to by their more common names. Thus, the Group IA elements are called the alkali metals, the Group IIA elements are called the alkaline earth metals, the Group VIIA elements are called the halogens, and the Group VIIIA elements are called the noble gases.

---

1) Author's Note: In 1986, the International Union of Pure and Applied Chemistry (IUPAC) proposed a new numbering system for the groups of the Periodic Table, consisting of the straightforward 1 through 18 from left to right. As this is being written, the controversy over the "best" way to number the groups remains unresolved. The system described above, while less straightforward than 1 through 18, retains certain advantages which will prove useful to students.

METALS AND NONMETALS:

The elements boron (B), silicon (Si), arsenic (As), tellurium (Te), and astatine (At) lie on a diagonal line in the Periodic Table. These elements and all elements which lie to the right of them on the Periodic Table are called nonmetals. Most of the elements which lie to the left of these on the Periodic Table are called metals. (The only exception is hydrogen, which is in Group IA but is a gas, not a metal!)

As you are probably aware, metals have certain properties in common which make them easy to recognize. Metals are shiny. In fact, the particular kind of shine common to most metals is so unique that it has been called the "metallic luster". Metals are malleable -- they can be molded into different shapes or drawn out into thin wires. This last property is especially useful, since metals are good conductors of heat and electricity. Most metals have high melting points and boiling points. (The best-known exception is mercury (Hg), which is a liquid at room temperature and is used in thermometers.)

Nonmetals are not quite so easy to characterize as a group, since their properties are so diverse. For example, the halogens (Group VIIA) include gases (fluorine and chlorine), a liquid (bromine), and a solid (iodine) among their ranks. Most nonmetals are non-conductors (sometimes called insulators) of heat and electricity. (A notable exception is carbon in its graphite form, which conducts electricity.) The solid nonmetals tend to be brittle, unlike the metals.

Some of the elements which lie near the "borderline" between metals and nonmetals on the Periodic Table have properties intermediate between those of metals and nonmetals. These elements are called metalloids or semi-metals, and they have been widely used in the semiconductor industry. Silicon, arsenic, tellurium, and germanium (Ge) are some of the elements which have these properties.

CHEMICAL PROPERTIES OF METALS AND NONMETALS:

Ions are formed when an atom gains or loses one or more electrons. As an example, consider a hydrogen atom, which has only one electron. Hydrogen atoms either lose their electrons to form H^+ ions, or gain electrons (one electron per atom) to form H^- ions. This is typical behavior for nonmetals, whose atoms may either lose or gain electrons to form either cations (positive ions) or anions (negative ions). Metals do not have this choice. Atoms of metals always react by losing electrons, forming positive ions.

The number of electrons lost by a metal atom in forming an ion can be predicted from the Periodic Table. For the main-group metals ("A" in the group number), the number of electrons lost in forming an ion is simply the number of the group. Thus, sodium atoms (Group IA) form Na^+ ions, calcium atoms (Group IIA) from Ca^{2+} ions, and aluminum atoms (Group IIIA) from Al^{3+} ions. Transition metals ("B" in the group number) also form cations, but their chemistry is more difficult to predict because many transition metals may form more than one cation. For example, chromium atoms (Group VIB) form Cr^{6+} and Cr^{3+} ions, among others.

The number of electrons gained by a nonmetal atom in forming an ion can also be predicted using the Periodic Table. Atoms of nonmetals tend to gain (8 - GN) electrons, where "GN" is the group number. The ions formed in this process have charges equal to (GN - 8). Thus, chlorine atoms (Group VIIA) gain one electron each (8 - 7 = 1) to form Cl^- ions (7 - 8 = -1). Oxygen atoms (Group VIA) gain two electrons each (8 - 6 = 2) to form O^{2-} ions (6 - 8 = -2). Nitrogen atoms (Group VA) gain three electrons each to form N^{3-} ions, and so on.

Problem: Recalling that the positive and negative charges in a chemical formula must cancel each other, write formulas for the products when hydrogen reacts with: a) Ca, b) Al, c) Cl. (Solutions: CaH_2, AlH_3, HCl.)

THE ORIGINS OF SOME ACIDS AND BASES:

Some covalent compounds form ions when dissolved in water. Examples include the covalent gases HCl and NH_3, as shown below:

$$HCl_{(g)} + H_2O_{(l)} \rightarrow H_3O^+_{(aq)} + Cl^-_{(aq)}$$

$$NH_{3\,(g)} + H_2O_{(l)} \rightarrow NH_4^+{}_{(aq)} + OH^-_{(aq)}$$

In 1903, the Swedish chemist Svante Arrhenius won the Nobel Prize in Chemistry for his studies of compounds such as HCl and NH_3 and their solutions. Arrhenius defined any substance which forms H_3O^+ ions (called hydronium ions) when dissolved in water as an acid, and any substance which forms OH^- ions (hydroxide ions) when dissolved in water as a base. By Arrhenius' definition, HCl is an acid (hydrochloric acid) and NH_3 (ammonia) is a base -- see the reactions above!

Some acids and bases can be formed by the reaction of various elements with oxygen and water vapor in the atmosphere. First, the element reacts with the oxygen to form a compound called an oxide. The oxide then reacts with the water vapor to form the acid or base.

Oxides of metals react with water to form bases. Examples include:

$$Na_2O_{(s)} + H_2O_{(l)} \rightarrow 2\ NaOH_{(aq)}$$

$$CaO_{(s)} + H_2O_{(l)} \rightarrow Ca(OH)_{2\,(aq)}$$

Oxides of nonmetals react with water to form acids. Examples include:

$$SO_{3\,(g)} + H_2O_{(l)} \rightarrow H_2SO_{4\,(l)}$$

$$N_2O_{5\,(g)} + H_2O_{(l)} \rightarrow 2\ HNO_{3\,(l)}$$

These last two reactions are important, because SO_3 and N_2O_5 are among the air pollutants emitted by some factories. When they react with water vapor in the atmosphere, the acids that form (H_2SO_4 = sulfuric acid, HNO_3 = nitric acid) return to earth in the form of acid rain. This causes severe environmental problems, such as fish kills and deterioration of forests.

RELATIVE STRENGTHS OF ACIDS AND THE PERIODIC TABLE:

When HCl reacts with H_2O, all of the HCl molecules dissociate (break up) to form H_3O^+ and Cl^- ions. Acids such as HCl, which are totally dissociated in water, are called strong acids. When HF reacts with H_2O, only about 3% of the HF molecules dissociate to form H_3O^+ and F^- ions -- the other 97% remain as HF molecules. Acids such as HF, which are only partially dissociated in water, are called weak acids. (Bases can also be strong or weak. One mole of NaOH dissolved in water will give one mole of OH^- ions, but one mole of NH_3 dissolved in water will only give about 0.01 mole of OH^- ions. Thus, NH_3 is a weak base.)

The Periodic Table can be used to determine which of two acids is the stronger acid. First, consider the formulas of the acids in question to determine whether they are binary acids or oxyacids. Binary acids are composed of hydrogen and one other element -- only two different elements are present in the formula. Oxyacids are composed of hydrogen, oxygen, and one other element -- a total of three different elements are present in the formula of an oxyacid.

The rule for determining the relative strengths of binary acids is that the closer the "other" element is to the lower right corner of the Periodic Table, the stronger the acid. HI is a stronger acid than HBr; I is closer to the lower right corner than Br is. Similarly, HBr is a stronger acid than H_2Se. (Note: Don't be fooled by the number of hydrogens present -- just use the rule!)

The rule for determining the relative strengths of oxyacids is that the closer the "other" element is to the upper right corner of the Periodic Table, the stronger the acid. H_2SO_4 is a stronger acid than H_2SeO_4; S is closer to the upper right corner than Se is. Similarly, $HClO_4$ is a stronger acid than H_2SO_4. (Don't let the number of hydrogens fool you!) The strength of an oxyacid does increase as the number of oxygens present increases. HNO_3 is stronger than HNO_2.

ELECTROMAGNETIC ENERGY:

To understand more about the chemical behavior of atoms, we need to learn more about the configuration of the electrons around an atom's nucleus. Scientists have approached this problem by exposing atoms to various forms of electromagnetic energy. Since the first instruments used to do these experiments were called spectroscopes, the study of atoms using electromagnetic energy is called spectroscopy.

Electromagnetic energy consists of electrical and magnetic forces whose intensities oscillate as time elapses. You've seen something like this if you've ever thrown a stone into a still pond of water. The "ripples" which spread out from the point where the stone hit the water are called traveling waves. The position of anything on the surface of the water (such as a leaf or an insect) oscillates (moves up and down repeatedly) as the traveling waves pass by it.

The frequency of a wave of electromagnetic energy is simply the number of oscillations it undergoes in one second. Frequency is represented by the Greek letter nu (ν), and has units of cycles per second (an oscillation is sometimes called a "cycle"), also known as sec^{-1} or Hertz. The wavelength of a wave of electromagnetic energy is just the distance that the wave travels during one cycle. Wavelength is represented by the Greek letter lambda (λ), and has units of length or distance. Meters can be used, but often a more convenient unit is the nanometer. (The prefix "nano-" means 10^{-9}. One meter = 10^9 nanometers.)

All electromagnetic energy travels at the speed of light, which is represented by the letter c and is equal to 3.00×10^8 meters per second. Hence, the product of the frequency and the wavelength of a beam of electromagnetic energy is always equal to the speed of light: 3.00×10^8 m/sec = c = $\lambda\nu$. (Bad Joke: First Physicist: "What's Nu?" Second Physicist: "C over Lambda!")

CONTINUOUS AND DISCONTINUOUS SPECTRA:

The many forms of electromagnetic energy make up a continuous spectrum -- all possible wavelengths and frequencies are represented below.

E.M. Energy	Wavelength, meters	Frequency, Hertz
Radio Waves	3×10^4 - 30	10^4 - 10^7
TV Waves	30 - 3	10^7 - 10^8
Radar	3 - 0.3	10^8 - 10^9
Microwaves	0.3 - 3×10^{-4}	10^9 - 10^{12}
Infrared	3×10^{-4} - 7×10^{-7}	10^{12} - 4×10^{14}
Visible Light	7×10^{-7} - 4×10^{-7}	4×10^{14} - 7×10^{14}
Ultraviolet	4×10^{-7} - 3×10^{-8}	7×10^{14} - 10^{16}
X Rays	3×10^{-8} - 3×10^{-11}	10^{16} - 10^{19}
Gamma Rays	3×10^{-11} - 3×10^{-16}	10^{19} - 10^{24}

A more common example of a continuous spectrum is the "rainbow" seen when white light is separated into its component colors by a prism. Each color "blends into" the next; there are no "gaps" in the spectrum. However, if a sample of hydrogen gas receives a high-energy electric spark, the H_2 molecules break apart into individual hydrogen atoms, which then get rid of their excess energy in the form of reddish-purple light. If the light emitted from these hydrogen atoms is passed through a prism, only the following wavelengths of visible light are present: 656 nm (red), 486 nm (green), 434 nm (indigo), and 410 nm (violet). The absence of the other wavelengths of visible light explains why the hydrogen spectrum is referred to as a discontinuous spectrum or a line spectrum -- there are "gaps" in the spectrum; only a few wavelengths are present. The hydrogen spectrum was discovered by Balmer in 1885. The spectra produced by other elements are also discontinuous. Scientists sought an explanation for this phenomenon.

PLANCK'S QUANTUM THEORY AND THE PHOTOELECTRIC EFFECT:

Hot objects glow. The hotter they get, the more the color of the glow moves toward the blue end of the visible spectrum. Hot objects glow red, hotter objects glow orange, really hot objects glow white, and the hottest stars glow bluish-white. Classical physics predicts that this trend should continue indefinitely, with extremely hot objects giving off ultraviolet light as their primary emission. They don't. This situation was called the "ultraviolet catastrophe", because classical physics failed to predict the trend correctly.

In 1900, the German physicist Max Planck found a solution to this problem. The solution was based on the idea that energy comes in small particles which Planck called photons or quanta, and that the energy of a quantum is related to its frequency. In equation form, Planck's quantum theory is: $E_{photon} = h\nu$. Here, E_{photon} is the energy of a photon in joules, ν is its frequency in sec^{-1}, and h is Planck's Constant, which has the value: $h = 6.63 \times 10^{-34}$ joule seconds. Planck was awarded the 1918 Nobel Prize in Physics for this work.

The quantum theory received some support in 1905 from another German physicist -- Albert Einstein. Einstein was studying the photoelectric effect, which was first observed in 1887 by the German physicist Heinrich Hertz. Hertz observed that shining electromagnetic energy on some metals caused them to lose electrons from their surfaces. Furthermore, the intensity of the energy used didn't matter -- what did matter was that the energy had to have a certain minimum frequency (called the threshhold frequency) for this effect to occur. Einstein explained this phenomenon using the quantum theory. If the energy of a quantum is proportional to its frequency, then only photons with a certain minimum frequency (or a greater frequency) will have enough energy to produce this effect. For this work (not the Theory of Relativity!), Einstein won a Physics Nobel Prize in 1921.

BOHR'S "SOLAR SYSTEM" MODEL OF THE ATOM:

In 1913, the Danish physicist Niels Bohr proposed a "solar system" model of the atom to try to explain the line spectrum obtained by the emission of light from excited hydrogen atoms. According to Bohr's model, electrons were restricted to certain orbits (also called energy levels) surrounding the atomic nucleus -- they could not exist between these orbits. (A good analogy to this situation is someone standing on a staircase. S/he can be standing on the first step, the second step, or the third step, but not on the "$1\frac{1}{2}$th" step or the "$2\frac{1}{4}$th" step -- no such step exists!) Bohr described the energies of the energy levels by the equation: $E_{el} = -k/n^2$, where E_{el} is the energy of an energy level (or of an electron in that energy level), k is a constant with a value of 2.18×10^{-18} J, and n is just the number of the energy level. The energy levels are consecutively numbered, with the closest orbit to the nucleus having n = 1, the next orbit out having n = 2, the next orbit having n = 3, the next having n = 4, and so on.

Bohr postulated that when an atom absorbs electromagnetic energy, it stores that energy, by moving its electrons from the ground state (the energy level where n = 1 and E_{el} is a minimum) to higher-energy excited states (energy levels where n = 2 or more and E_{el} is greater than -k). When an atom emits electromagnetic energy, the reverse process occurs -- electrons in high-energy excited states move into lower-energy states.

Bohr's model also supports Planck's quantum theory. If the electrons in an atom can exist only in certain energy levels, and if they move from one level to another when the atom absorbs energy, then it must be true that the atom can only absorb certain amounts of energy. This is just another way of saying that energy is quantized -- that is, it comes in discrete "lumps" called quanta.

Bohr was awarded the 1922 Nobel Prize in Physics for this work.

USING BOHR'S MODEL TO PREDICT THE HYDROGEN EMISSION SPECTRUM:

The major significance of the Bohr model is that it allows for the prediction of the discontinuous spectrum of visible light emitted by excited atoms of hydrogen. The hydrogen spectrum consists of visible light with the following wavelengths: 656 nm, 486 nm, 434 nm, and 410 nm. Does the model predict this?

Problem: What is the wavelength of the photon of light emitted when an electron moves from the $n = 3$ level to the $n = 2$ level in the hydrogen atom?

Solution: When an electron moves from one energy level to another, the energy of the electron changes. The Law of Conservation of Energy tells us that the electron's change in energy must be equal to the energy of the photon of light that is emitted. Thus, we need to find the energies of the energy levels involved -- here, the $n = 3$ and $n = 2$ energy levels. Given that $E_{el} = -k/n^2$ and knowing that $k = 2.18 \times 10^{-18}$ joules, we can calculate the following:

$$E_3 = -2.18 \times 10^{-18} \text{ J}/3^2 = -2.18 \times 10^{-18} \text{ J}/9 = -2.42 \times 10^{-19} \text{ joules.}$$

$$E_2 = -2.18 \times 10^{-18} \text{ J}/2^2 = -2.18 \times 10^{-18} \text{ J}/4 = -5.45 \times 10^{-19} \text{ joules.}$$

Now, simply subtracting these two values gives us the energy of the photon:

$$E_{photon} = E_3 - E_2 = (-2.42 \times 10^{-19} \text{ joules}) - (-5.45 \times 10^{-19} \text{ joules})$$
$$= 3.03 \times 10^{-19} \text{ joules.}$$

Planck's equation ($E_{photon} = h\nu$) can be used to convert this into a frequency:

$$\nu = E_{photon}/h = (3.03 \times 10^{-19} \text{ joules})/(6.63 \times 10^{-34} \text{ joule sec})$$
$$= 4.57 \times 10^{14} \text{ sec}^{-1}.$$

Finally, the wavelength can be found using the $c = \lambda\nu$ relationship:

$$\lambda = c/\nu = (3.00 \times 10^{8} \text{ meters/sec})/(4.57 \times 10^{14} \text{ sec}^{-1})$$
$$= 6.56 \times 10^{-7} \text{ meters} = 656 \times 10^{-9} \text{ meters} = \underline{656 \text{ nanometers!}}$$

Notice that this is one of the wavelengths observed in the hydrogen spectrum! The other wavelengths above can be calculated in a similar way. See if you can do it!

THE WAVE NATURE OF MATTER:

Bohr's "solar system" model of the atom turned out to be useful <u>only</u> for predicting the emission spectrum of <u>hydrogen</u>. For other elements, the Bohr model failed miserably, so another model was needed for more complex atoms.

In 1924, a French graduate student in physics, Louis deBroglie, combined Planck's equation ($E = h\nu$) and Einstein's famous relationship between mass and energy ($E = mc^2$) into one equation: $h\nu = mc^2$. Dividing both sides of this equation by $mc\nu$ gives us a new equation: $\frac{h}{mc} = \frac{c}{\nu} = \lambda$. (since $c = \lambda\nu$) deBroglie's equation, for which he won the 1929 Nobel Prize in Physics, shows that <u>matter</u> (represented by m in the equation) has <u>wave</u> properties (represented by λ).

Following the publication of deBroglie's (four-page!) Ph.D. thesis which included the above hypothesis, numerous experiments supported the idea that <u>wave</u> properties are exhibited by some forms of <u>matter</u>. For example, in 1927, physicists Clinton Davisson and George Thomson (son of J. J. Thomson, who discovered electrons!) found that a beam of electrons passing through a crystal was scattered in such a way as to produce a "diffraction pattern" -- a series of alternating bright spots and dark spots which occurs when <u>waves</u> reinforce each other or cancel each other out, respectively. (The analogy of two stones thrown into a still pond may be useful in thinking about what a "diffraction pattern" is. As the "ripples" spread out from each stone, they will eventually overlap each other. The pattern of waves that forms when they overlap is the "diffraction pattern" for these waves.) Since electrons behave like <u>waves</u> in this experiment, scientists began to look for <u>wave</u> models to describe the behavior of electrons in atoms. Davisson and Thomson shared the 1937 Nobel Prize in Physics for this work. (Ironically, J. J. Thomson showed that electrons were <u>particles</u>, while his son George showed that electrons were <u>waves</u>! Both men deserved their Nobel Prizes.)

WAVE EQUATIONS FOR ELECTRONS:

Beams of electrons behave like traveling waves (that is, like ripples on a still pond when a stone is thrown into it), but electrons in atoms aren't traveling in a straight line. Instead, they tend to stay relatively close to the atomic nucleus. Electrons in atoms are thought to behave more like standing waves. A common example of a standing wave is a vibrating guitar string. The string obviously oscillates back and forth, but it doesn't travel anywhere as long as the guitar itself stands still. Different wave patterns can be produced by holding down one or more points along the guitar string before plucking it, but the wave patterns are limited by the fact that the ends of the guitar string are already tied to the guitar. The wave equation which describes the wavelength of a standing wave is: $\lambda = 2L/n$, where λ is the wavelength, L is the length of the guitar string, and n is a number related to the number of nodes present in the standing wave. (A node is just a part of a standing wave that doesn't move -- in the guitar string, a part that's held down.) The point is that only certain values of the wavelength are possible for a standing wave -- just as in the Bohr model of the atom, only certain orbits were possible! In each case, n may have the values 1, 2, 3, 4, etc., but no values between these numbers are allowed.

In 1926, the Austrian physicist Erwin Schrödinger worked out the wave equations which describe the probability of finding electrons at various locations in the atom. We can only discuss the probability of finding an electron at a specific location, because any attempt to measure its exact location causes it to move. This was discovered in 1926 by the German physicist Werner Heisenberg, who stated the Heisenberg Uncertainty Principle this way: "It is impossible to know both the position and the momentum of a small particle simultaneously." Both Heisenberg (1932) and Schrödinger (1933) received Nobel Prizes in Physics.

ATOMIC ORBITALS:

The solutions to Schrödinger's wave equations for electrons in atoms are very complex, but they can be graphed to see what they look like. Actually, drawing a graph on paper is somewhat misleading -- paper is two-dimensional, but the solutions to the Schrödinger equations represent three-dimensional regions called orbitals. An orbital is simply a region in three-dimensional space in which electrons in atoms are most likely to be found. Orbitals are centered at the nucleus of an atom. They are called "orbitals" because they are similar to (but not the same as!) the "orbits" for electrons in the Bohr model of the atom. For example, just as the different "orbits" in the Bohr model had different energies, different orbitals also have different energies. When an atom absorbs energy or emits energy, the atom's electrons move from their original orbitals to other orbitals with different energies. However, since orbitals are just the graphs of wave equations, they have some properties similar to those of waves -- particularly, standing waves. For example, orbitals have different shapes, which are designated by different letters and tell us where electrons are most probably located. s orbitals are spherical, with the atomic nucleus located at the center of the sphere. s orbitals contain no nodes. p orbitals are shaped like "dumbbells". There are three kinds of p orbitals, arranged at 90^{0} angles to each other along the x, y, and z axes. p orbitals contain one node each. d orbitals come in groups of five. Four of them look like "four-leaf clovers"; the other one looks like a p orbital inside a "doughnut". d orbitals have two nodes each. f orbitals come in groups of seven. They are by far the most exotic-looking: three of them look like p orbitals inside two "doughnuts", and the other four look like "eight-leaf clovers", if you can imagine such a thing! f orbitals have three nodes. (We won't use f orbitals often, but we will use s, p, and d orbitals.)

QUANTUM NUMBERS:

The equation for the energy of an electron in the Bohr model of the atom is: $E_{el} = -k/n^2$, where n is a positive integer (1, 2, 3, 4, etc.) that represents the energy level in which the electron is located. The Bohr model worked well for hydrogen atoms, which only contain one electron each. However, a more complex equation (Schrödinger's) is needed to describe more complex atoms. Numbers such as n in the equation above are called quantum numbers, because they are quantized -- that is, they are only allowed to have certain values. The complete solution to Schrödinger's equation for complex atoms (those with more than one electron) requires four quantum numbers, which are described below.

The principal quantum number (symbolized by n) is similar to the "n" in the Bohr equation. It must be a positive integer (1, 2, 3, 4, etc.). As the value of n increases, the energy of the electron increases, and the distance of the electron from the atomic nucleus increases. (Just like the Bohr model!)

The azimuthal quantum number (symbolized by l) ranges in value from zero up to $n-1$, by integers: $l = 0, 1, 2, 3, \ldots, n-1$. The value of l designates the shape of the orbital in which the electron is located: $l = 0$ for an s orbital, $l = 1$ for a p orbital, $l = 2$ for a d orbital, and $l = 3$ for an f orbital.

The magnetic quantum number (symbolized by m) ranges in value from $-l$ to $+l$, by integers. For example, if $l = 2$ (a d orbital), then the allowed values of m are -2, -1, 0, +1, and +2 -- a total of five different values of m, which tells us that there are five different d orbitals.

The spin quantum number (symbolized by s) has only two possible values: $s = +1/2$ or $s = -1/2$. Both the electron and the atomic nucleus are surrounded by magnetic fields, and the value of s tells us whether these magnetic fields point in the same direction ("spin up") or in opposite directions ("spin down").

THE PAULI EXCLUSION PRINCIPLE:

In 1926, the Austrian physicist Wolfgang Pauli discovered that no two electrons in the same atom can have the same values for all four quantum numbers. This axiom, often called the Pauli Exclusion Principle, is useful because it allows us to use sets of quantum numbers like "zip codes" for electrons in atoms. No two cities have the same zip code, and no two electrons have the same "address" in the atom. Notice that since the first three quantum numbers (n, l, and m) refer to orbitals, and only the spin quantum number (s) refers specifically to electrons, a corollary of the Pauli Exclusion Principle is that there can be a maximum of two electrons in any one orbital. Furthermore, these two electrons must be "spin-paired" -- that is, one of them must be "spin up" ($s = +1/2$) and the other one must be "spin down" ($s = -1/2$). Single electrons in orbitals are sometimes called "unpaired" electrons as a result. Pauli won the 1945 Nobel Prize in Physics for this discovery.

All of the orbitals which have the same value for the n quantum number are referred to as a shell, after the spherical, "shell-like" orbits of the Bohr model of the atom. Notice that in the first shell ($n = 1$), only one orbital is present (the 1s orbital), so that the maximum number of electrons that can be present in the first shell is two. In the second shell ($n = 2$), four orbitals are present (the 2s orbital and the three 2p orbitals), so a maximum of eight electrons may be present in the second shell. In the third shell ($n = 3$), nine orbitals are present (3s, three 3p, and five 3d orbitals), so at most eighteen electrons may be present. By similar logic, thirty-two electrons is the most that can be present in the fourth shell. Now, look at the Periodic Table. There are two elements in the first row, eight elements each in Rows 2 and 3, eighteen in Rows 4 and 5, and thirty-two in rows 6 and 7 -- just as Pauli's principle predicts!

THE AUFBAU PRINCIPLE:

The Aufbau principle ("Aufbau" comes from German words which mean "to build up") states that electrons in atoms will occupy the orbitals with the lowest energy first, and only after these orbitals are filled will electrons occupy orbitals of higher energy. The energies of orbitals increase as the values of the first two quantum numbers (n and l) increase. Thus, the 1s orbital is lowest in energy, the 2s orbital is next lowest, then the 2p orbitals, the 3s orbital, the 3p orbitals, and so on in the following sequence: 4s, 3d, 4p, 5s, 4d, 5p, 6s, 4f, 5d, 6p, 7s, 5f, 6d, 7p. An easy way to remember this sequence is to construct a "Pascal's Christmas Tree"[1], which is illustrated at the right. Draw an eight-layer "Christmas tree" with one "bulb" each in rows 1 and 2, two "bulbs" each in rows 3 and 4, three "bulbs" each in rows 5 and 6, and four "bulbs" each in rows 7 and 8. Write an "s" in the right-hand bulb in each row, a "p" to the left of each "s", a "d" to the left of each "p", and an "f" to the left of each "d". Number the "s" bulbs from 1 to 8 (top to bottom), the "p" bulbs from 2 to 7, the "d" bulbs from 3 to 6, and the "f" bulbs from 4 to 5. Simply reading the rows across, from left to right, will give the above sequence of orbitals.

```
          1s
          2s
       2p  3s
       3p  4s
    3d  4p  5s
    4d  5p  6s
4f  5d  6p  7s
5f  6d  7p  8s
          !!
```

All of the orbitals which have the same values for both the n and l quantum numbers are referred to as a subshell. All of the orbitals in a subshell have the same energy, and are referred to as degenerate orbitals. When electrons are placed in a set of degenerate orbitals, they tend to "spread out" as much as possible -- one electron per orbital unless there are so many electrons that they must "pair up". This is known as Hund's Rule, sometimes called the "bus-seat" rule, because strangers boarding buses don't sit next to anyone unless they must!

1) Darsey, J. A., J. Chem. Educ. 65, 1036 (1988).

ELECTRON CONFIGURATIONS IN ATOMS:

Problem: What is the configuration of electrons in an oxygen atom?

Solution: Oxygen's atomic number is 8, so a neutral oxygen atom will contain eight electrons. The 1s orbital is lowest in energy, so we fill it first by placing two electrons in it. The next two electrons go into the 2s orbital, since it is next lowest in energy. Next come the three degenerate 2p orbitals, but we only have four electrons remaining. At this point, Hund's Rule says that we should place one electron in each of the three 2p orbitals, followed by placing the last electron in any of the 2p orbitals to complete an electron pair. The resulting electron configuration is shown in the energy-level diagram at the right, but a more convenient "shorthand" way of representing the electron configuration of an atom is to simply list the subshells which contain electrons and write the number of electrons in each subshell as superscripts. By this method, oxygen's electron configuration is: $1s^2 2s^2 2p^4$.

2p " ' '
2s "
1s "

Problem: What is the electron configuration in a tungsten atom?

Solution: Tungsten's atomic number is 74, so we need to place 74 electrons in orbitals. An energy-level diagram like the one above for this many electrons would be a tedious thing to draw, so we'll use the "shorthand" notation instead. All we need to do is remember the order in which the subshells fill up, and if we need help with that, we can make a "Pascal's Christmas Tree"[1] (see the previous page). Then, if we know how many electrons can go into each subshell (s = 2, p = 6, d = 10, f = 14), all we have to do is count to 74 electrons! The correct result is: $1s^2 2s^2 2p^6 3s^2 3p^6 4s^2 3d^{10} 4p^6 5s^2 4d^{10} 5p^6 6s^2 4f^{14} 5d^4$. This method will usually give the correct electron configuration, although there are some exceptions to the general rule (most of them fifth-row transition metals).

---

1) Darsey, J. A., J. Chem. Educ. 65, 1036 (1988).

ELECTRON CONFIGURATIONS AND THE PERIODIC TABLE:

The valence shell of an atom is the shell farthest from the nucleus which contains at least one electron. Many of the chemical properties of an atom are based on the number of valence electrons (electrons in the valence shell) that the atom possesses.

Problem: How many valence electrons does an atom of sulfur have?

Solution: Following the usual rules for determining electron configurations, we would obtain the following electron configuration for sulfur (atomic number = 16): $1s^2 2s^2 2p^6 3s^2 3p^4$. At this point, we can easily see that the valence shell is the third shell (the largest value of n written above is 3), and that this shell contains six electrons (two in the 3s subshell and four in the 3p subshell). Hence, there are six valence electrons present in an atom of sulfur. However, this method would be tedious for atoms with large atomic numbers. There is a much easier way to obtain the above information -- use the Periodic Table! Sulfur is in the third row and in Group VIA, and it has six valence electrons in the third shell. In general, for any of the "main-group" elements ("A" groups), the row number is the same as the value of n for the valence shell, and the group number is the number of valence electrons present in the atom.

Another way to write the above electron configuration is: $[Ne]3s^2 3p^4$. This "noble gas" form for writing electron configurations uses the symbol for a noble gas as "shorthand" for that gas ([Ne] = neon = $1s^2 2s^2 2p^6$) and emphasizes only the valence electrons by writing them outside the brackets.

Problem: What is the electron configuration for calcium (Ca) using the "noble gas" notation?

Solution: Calcium is in Row 4 and Group IIA. The noble gas with a filled third shell is argon, so the electron configuration is simply: $[Ar]4s^2$.

LEWIS SYMBOLS AND THE OCTET RULE:

When metals react with nonmetals to form ionic compounds, electrons are transferred *from* the atoms of the metal *to* the atoms of the nonmetal. The cations and anions that result are then attracted to each other. This attraction is called an *ionic bond*. One way to keep track of the transfer of electrons that occurs in reactions like this is to use *Lewis symbols*, which were developed by the American chemist G. N. Lewis. A Lewis symbol is simply the chemical symbol for an element surrounded by *dots* which represent the *valence electrons* in the atoms of the element. Some examples of Lewis symbols are shown below.

Na = $[Ne]3s^1$ = Na· P = $[Ne]3s^2 3p^3$ = :P· (one dot above, one dot below)

Cl = $[Ne]3s^2 3p^5$ = :Cl· (two dots above, two dots below) Ar = $[Ne]3s^2 3p^6$ = :Ar: (two dots above, two dots below)

Note that the arrangement of the dots corresponds to the arrangement of the electrons in the valence shell in each case. Lewis symbols are sometimes called "electron dot" symbols, for obvious reasons.

An example of the use of Lewis symbols to illustrate a reaction in which electrons are transferred from one atom to another is shown below by the reaction of a calcium atom (Ca = $[Ar]4s^2$) with a sulfur atom (S = $[Ne]3s^2 3p^4$):

$$\text{Ca:} + \text{:}\dot{\underset{\cdot}{\text{S}}}\text{:} \longrightarrow \text{Ca}^{2+} + \text{:}\ddot{\underset{\cdot\cdot}{\text{S}}}\text{:}^{2-}$$

Notice that the calcium atom *loses* two electrons, so that the Ca^{2+} ion has the same electron configuration as an argon atom -- that is, [Ar]. Notice also that the sulfur atom *gains* two electrons, so that the S^{2-} ion also has the [Ar] configuration. The atoms of the main-group elements generally tend to gain or lose electrons until the ions formed by this process have the same electron configuration as the atoms of a *noble gas*. This is sometimes referred to as the *octet rule*, since the products of these reactions are usually ions or atoms with *eight* valence electrons. (An "octet" is *eight* of something.)

COVALENT BONDS:

Covalent compounds are made of *molecules*. The atoms in molecules are connected by *covalent bonds*. The word "co-valent" implies that these bonds are made of electrons which occupy the *valence* shells of *two* atoms simultaneously -- that is, that the electrons in a covalent bond are "shared" between two atoms.

In order to understand this better, consider two atoms, X and Y. Each atom has *one* unpaired electron. As the atoms move closer together, each unpaired electron begins to be attracted to the nucleus of the *opposite* atom, as well as to its own atom's nucleus. The covalent bond is fully formed when each electron is equally attracted to *each* of the two nuclei. An equation for this is shown below:

$$X\cdot \;+\; \cdot Y \longrightarrow X{:}Y$$

Covalent bonds can be represented by *dots* (for individual electrons) or by *dashes* (for *pairs* of electrons). For example, a molecule of H_2 could be represented by either H:H or H-H. Notice that the dots or dashes are drawn *between* the connected atoms, since the electrons in covalent bonds are located (roughly) between the two atomic nuclei that they join. Representations of covalent molecules like H:H or H-H are called *Lewis structures*, after the American chemist G. N. Lewis.

The covalent bond in H_2 is called a *single* bond, since it is composed of only *one* pair of electrons. *Multiple* bonds, consisting of two or more pairs of electrons, are also known to exist. Examples include the two *double* bonds in each molecule of CO_2 (:Ö::C::Ö: or :Ö=C=Ö:) and the *triple* bond in each molecule of N_2 (:N:::N: or :N≡N:).

Since electrons are "shared" -- not transferred -- in covalent bonds, the "octet rule", which states that atoms tend to either gain or lose electrons until they have *eight* valence electrons each, needs to be modified somewhat in order to predict what the electrons in covalent molecules will do.

THE MODIFIED OCTET RULE:

The modified octet rule states that the atoms of the "main-group" elements tend to gain or lose electrons until they each have eight electrons in their valence shells, even if some of those eight electrons are "shared" with other atoms. Consider the Lewis structures of Cl_2, HCN, and CO_2 shown below:

:C̤̈l:C̤̈l: or :C̤̈l-C̤̈l: H:C:::N: or H-C≡N: :Ö̤::C::Ö̤: or :Ö̤=C=Ö̤:

Notice that each atom (except H) in the above Lewis structures is surrounded by a total of eight electrons, whether they are used in covalent bonds or not. (Pairs of electrons which are not part of a covalent bond are usually called lone pairs.) In determining this, notice that electrons in bonds are counted twice -- once for each atom. For example, each chlorine atom in Cl_2 is surrounded by eight valence electrons, despite the fact that there are only fourteen valence electrons present in the entire molecule! The two electrons in the bond are part of both "octets".

Notice that we said "except H" above. Hydrogen is an exception to the octet rule; hydrogen atoms only require two electrons in each of their valence shells. (This should make sense: the valence shell of a hydrogen atom is the first shell, and the first shell (n = 1) can hold a maximum of two electrons.) Other exceptions include elements in Group IIA (beryllium (Be) only requires four valence electrons) and Group IIIA (boron (B) only requires six valence electrons). In addition, elements which lie below the second row of the Periodic Table may include more than eight electrons in their valence shells if necessary. For example, the sulfur atom in the SF_6 molecule is surrounded by twelve electrons.

In general, the number of covalent bonds connected to the atoms of a "main-group" element is (8 - GN), where "GN" is the group number of the element. Thus, carbon atoms have four bonds, nitrogen atoms have three, oxygen atoms have two, and halogen atoms have one. (Check the Lewis structures above!)

HOW TO DRAW LEWIS STRUCTURES:

The correct Lewis structures for many molecules and polyatomic ions can be easily drawn simply by following the rules outlined below.

Step # 1: Count the total number of valence electrons for each atom in the molecule. Add these numbers together, add one for each negative charge present on the molecule or ion, and subtract one for each positive charge present.

Step # 2: Draw the skeletal structure (the basic arrangement of the atoms) of the molecule or ion. If there is a unique atom (that is, only one atom of a particular element), place that atom in the center of the molecule or ion.

Step # 3: Connect adjacent atoms with single bonds.

Step # 4: Surround all atoms except the central atom with enough lone pairs to complete their octets. (Remember the exceptions to the octet rule!)

Step # 5: Count the number of valence electrons in your drawing. (Each single bond or lone pair = two electrons.) Subtract the total from the total you got in Step # 1. If any valence electrons remain, place them on the central atom in the form of lone pairs (whenever possible).

Step # 6: If the central atom does not have a complete "octet" of electrons, complete its "octet" by moving lone pairs on adjacent atoms toward the central atom, forming multiple bonds until the central atom's "octet" is complete.

Step # 7: Check for violations of the "8 - GN" rule, which predicts the number of covalent bonds that should be attached to each atom.

The following is the result of each step above for a molecule of HNO_2.

Step # 1: H = 1, N = 5, O = 6, O = 6. 1 + 5 + 6 + 6 = 18 electrons.

Step # 2: H O N O (N is the central atom, since H can only have one bond.)

```
                          ..   ..                   ..  ..  ..
Step # 3:  H-O-N-O   Step # 4:  H-O-N-O:   Step # 5:  H-O-N-O:   (18 - 16 = 2 e⁻'s.)
                          ..   ..                   ..      ..
                       ..  ..  ..
Steps # 6 and # 7:   H-O-N=O:   (H-O=N-O: violates the "8 - GN" rule.)
                       ..                ..  ..  ..
```

DRAWING LEWIS STRUCTURES:

<u>Problem</u>: Draw the correct Lewis structure for the ClF_3 molecule.

<u>Solution</u>: Cl = 7, F = 7. 7 + 7 + 7 + 7 = <u>28</u> valence electrons. Place the unique atom (Cl) in the center, then surround it with three fluorine atoms. Filling in the single bonds (three) and lone pairs around the fluorine atoms (three lone pairs each, for a total of nine) requires the use of a total of <u>24</u> electrons. 28 - 24 = <u>4</u> electrons which must be placed on the central Cl atom. The final Lewis structure is shown at the right. Note that the Cl atom is surrounded by <u>ten</u> electrons. This violation of the octet rule is permissible here, since Cl is below the second row of the Periodic Table.

```
 ¨  ¨¨  ¨
:F:Cl:F:
 ¨  ¨   ¨
   :F:
    ¨
```

<u>Problem</u>: Draw the correct Lewis structure for the CS_2 molecule.

<u>Solution</u>: C = 4, S = 6. 4 + 6 + 6 = <u>16</u> valence electrons. Place the unique atom (C) in the center, then place the sulfur atoms on each side. Filling in the single bonds (two) and lone pairs around the sulfur atoms (three lone pairs each, for a total of six) requires the use of a total of <u>16</u> electrons. No electrons remain to be added to the central atom (16 - 16 = 0). However, the carbon atom is only surrounded by <u>four</u> electrons -- it needs four more to complete its octet. This problem is solved by moving one lone pair from each sulfur atom toward the central carbon atom, forming <u>two double bonds</u>: :S̈=C=S̈:

<u>Problem</u>: Draw the correct Lewis structure for the <u>cyanide ion (CN^-)</u>.

<u>Solution</u>: C = 4, N = 5, negative charge = 1. 4 + 5 + 1 = <u>10</u> valence electrons. Simply connect the two atoms with a single bond, then complete the octet of either atom with three lone pairs. This requires <u>eight</u> electrons, which leaves <u>two</u> electrons to go onto the other atom (10 - 8 = 2). At this point, one atom will only have <u>four</u> electrons surrounding it. To complete this atom's octet, move <u>two</u> lone pairs from the other atom to form a <u>triple bond</u>: :C≡N:

RESONANCE STRUCTURES:

<u>Problem</u>: Draw the correct Lewis structure for <u>ozone (O_3)</u>.

<u>Solution</u>: Oxygen is in Group VIA. 3 x 6 = <u>18</u> valence electrons. Since there is no unique atom here, place <u>any</u> atom in the center and surround it with the other two -- that is, arrange the three oxygen atoms in a straight line. Filling in the single bonds (two) and lone pairs around the <u>outer</u> oxygen atoms (three lone pairs each, for a total of six) requires the use of a total of <u>16</u> electrons. 18 - 16 = <u>2</u> electrons which must be placed on the central oxygen atom. This leaves us with the structure $:\underset{..}{\overset{..}{O}}-\overset{..}{O}-\underset{..}{\overset{..}{O}}:$, in which the central oxygen atom only has <u>six</u> valence electrons surrounding it. To complete the octet, one lone pair should be moved in from one of the outer oxygen atoms to form a double bond. It makes no difference <u>which</u> of the two outer oxygen atoms donates a lone pair to the central atom, since they are essentially equivalent. Therefore, the final Lewis structure of ozone could be written as either $:\overset{..}{O}=\overset{..}{O}-\underset{..}{\overset{..}{O}}:$ or as $:\underset{..}{\overset{..}{O}}-\overset{..}{O}=\overset{..}{O}:$ -- either one is a reasonable possibility. When more than one reasonable Lewis structure can be drawn for a molecule, the structures are called <u>resonance structures</u>. Ozone does not really "resonate" between these two structures. The <u>actual</u> structure of ozone is a <u>hybrid</u> of the two structures. This is symbolized as shown below.

$$:\overset{..}{O}=\overset{..}{O}-\underset{..}{\overset{..}{O}}: \longleftrightarrow :\underset{..}{\overset{..}{O}}-\overset{..}{O}=\overset{..}{O}: \quad = \quad :\overset{..}{O}\overset{---}{=}\overset{..}{O}\overset{---}{=}\overset{..}{O}: \text{ (actual structure)}$$

The double-headed arrow above is used to indicate resonance structures. The dotted line between the oxygen atoms in the hybrid structure indicates that the bond it represents is not a full covalent bond, but a sort of "half-bond" -- the two electrons in question are distributed over <u>all three</u> oxygen atoms. This may sound strange, but it is important to realize that the actual structure of ozone is <u>best</u> represented by the <u>resonance hybrid</u> structure above. The resonance structures just help us to visualize the actual structure a little more easily.

FORMAL CHARGES AND COORDINATE COVALENT BONDS:

An atom which carries an electrical charge will violate the "8 - GN" rule, which is used to predict the number of covalent bonds normally attached to that atom. Electrical charges which are located on atoms are sometimes called formal charges. For an example of this, consider the hydronium ion (H_3O^+), for which the Lewis structure is drawn below:

```
  ..               ..
H:O:  +  H+  -->  H:O:H
  ..               ..
  H                H
```

Since oxygen is in Group IIA, a normal oxygen atom should only have two covalent bonds. (8 - 6 = 2.) The oxygen atom in the hydronium ion has three covalent bonds. This is acceptable as long as we realize that the oxygen atom in the hydronium ion also carries the ion's positive electrical charge. Another example is the BH_4^- ion, for which the Lewis structure is shown below:

```
  H                  H
  ..                 ..
H:B  +  :H-  -->   H:B:H
  ..                 ..
  H                  H
```

Boron is an exception to the "8 - GN" rule. Boron is in Group IIIA, and its atoms normally have three covalent bonds. In the BH_4^- ion, the boron atom has four covalent bonds. Again, this indicates that the boron atom is the atom that carries the ion's negative electrical charge.

The polyatomic ions above each contain a coordinate covalent bond, which is just like any other covalent bond except for the fact that the two electrons in the bond came originally from the same atom or ion. A simple way to determine the formal charge on an atom is to use the equation: $F.C. = GN - U - \frac{S}{2}$, where GN is the atom's group number, U is the number of unshared electrons around the atom, and S is the number of shared electrons surrounding the atom.

VALENCE SHELL ELECTRON PAIR REPULSION THEORY:

Electrons, being negatively charged, tend to repel each other. This property of electrons is the basis of the Valence Shell Electron Pair Repulsion (VSEPR) Theory, which states that the electron pairs in the valence shell of an atom will tend to arrange themselves in such a way as to be as far apart from each other as possible. The shapes of covalent molecules can be predicted from their Lewis structures by applying the VSEPR theory to the central atom in the molecule. In this context, "electron pairs" refers to either lone pairs or covalent bonds, regardless of whether the covalent bonds are single bonds or multiple bonds.

An atom with two pairs of electrons in its valence shell would have, as its lowest energy state, the electron pairs located on opposite sides of the atom. This is called a linear geometry, since the electron pairs and the atom all lie on one line. If three electron pairs are present in an atom's valence shell, they will be arranged in a triangular pattern around the atom. Four electron pairs produce a tetrahedral geometry. (A tetrahedron is a four-sided figure which looks like a pyramid with a triangular base. If the atom is located at the center of a tetrahedron, the electron pairs would point toward its corners.) Five pairs of electrons produce a trigonal bipyramidal geometry. (A trigonal bipyramid is a six-sided figure which looks like two pyramids stuck together at one triangular face. If the atom is located at the center of a trigonal bipyramid, the electron pairs would point toward its corners.) Six pairs of electrons produce a geometry called octahedral. (An octahedron is an eight-sided figure which looks like two square-based pyramids stuck together at the square base. If the atom is located at the center of an octahedron, the electron pairs would point toward its corners.) These five basic geometries are sufficient to describe the shapes of a large number of covalent molecules.

PREDICTING THE SHAPES OF MOLECULES USING VSEPR THEORY:

Problem: What is the geometry of the CO_2 molecule?

Solution: The Lewis structure of CO_2 is shown at the right. :Ö=C=Ö:
The carbon atom is the central atom, and it is surrounded by only *two* "electron pairs" (each double bond counts as only *one* "electron pair" when using the VSEPR theory). The molecule is *linear*, as indicated in the Lewis structure. Notice that this geometry implies an O-C-O *bond angle* of 180°.

Problem: What is the geometry of the BH_3 molecule?

Solution: The Lewis structure of BH_3 is shown at the right.

```
  H
  ¨
H:B:H
```

The boron atom is the central atom, and it is surrounded by only *three* electron pairs (three single bonds), so the molecule has a *triangular geometry*. A better drawing of the BH_3 molecule is shown at the right. Notice that this geometry implies three equal H-B-H bond angles of 120°.

```
   .H
H:B:
   .H
```

Problem: What is the geometry of the CH_4 molecule?

Solution: The Lewis structure of CH_4 is shown at the right.
From the Lewis structure, a square geometry with 90° H-C-H bond angles would be predicted, but the actual geometry of the CH_4 molecule is the *tetrahedral geometry*, with H-C-H bond angles of *109.5°*. One feature of the VSEPR theory is that the bond angles should always be as large as possible.

```
  H
  ¨
H:C:H
  ¨
  H
```

Problem: What is the geometry of the H_2O molecule?

Solution: The Lewis structure of H_2O is shown at the right.

```
  H
  ¨
H:O:
  ¨
```

Like CH_4, H_2O has *four* electron pairs surrounding the central atom. Unlike CH_4, two of those electron pairs are lone pairs. The lone pairs repel the single bonds, giving the H_2O molecule a *bent* geometry (not linear like CO_2!). The H-O-H bond angle in H_2O is approximately 105°, due to the fact that lone pairs repel other electron pairs a little bit more strongly than covalent bonds do.

A SUMMARY OF MOLECULAR GEOMETRIES:

The table below summarizes the terms used to describe the geometries of molecules with various numbers of electron pairs surrounding the central atom.

Total Electron Pairs Around Central Atom	Total Lone Pairs Around Central Atom	Term Used To Describe Geometry Of Molecule	One Example Of A Molecule With This Geometry
2	0	Linear	CO_2
	1	Linear	N_2
3	0	Triangular	BH_3
	1	Bent	SO_2
4	0	Tetrahedral	CH_4
	1	Pyramidal	NH_3
	2	Bent	H_2O
	3	Linear	HF
5	0	Trigonal Bipyramidal	PCl_5
	1	"Seesaw"-Shaped	SF_4
	2	"T"-Shaped	ClF_3
	3	Linear	XeF_2
6	0	Octahedral	SF_6
	1	Square Pyramidal	BrF_5
	2	Square	XeF_4

Bond angles for each of the five basic geometries:

Linear	180^0
Triangular	120^0
Tetrahedral	109.5^0
Trigonal Bipyramidal	90^0 and 120^0
Octahedral	90^0

POLAR COVALENT BONDS AND ELECTRONEGATIVITY:

In some covalent bonds, the electrons are distributed symmetrically between the connected atoms. In other covalent bonds, this is not the case -- one atom has a greater share of the electrons in the bond than the other atom does. If the electrons in a covalent bond are more attracted to one atom in the bond than to the other atom, then the atom which has the greater share of the electrons will also have a partial negative charge. The atom which has the lesser share of the electrons will be left with a partial positive charge. The covalent bond thus has two "opposite ends", or poles. It is therefore called a polar covalent bond. The partial positive and partial negative charges in a polar covalent bond are represented by "δ+" and "δ-" respectively. ("δ" is the lower-case Greek letter delta.) These charges are not as strong as the full positive ("+") and negative ("-") charges present in ionic bonds, but they are measurable.

An example of a molecule which contains a polar covalent bond is HF. The fluorine atom draws the electrons in the covalent bond toward itself and thus acquires a δ- charge, leaving the hydrogen atom with a δ+ charge. Atoms such as fluorine, which tend to draw the electrons in covalent bonds toward themselves, are called electronegative. Fluorine atoms are the most electronegative, oxygen atoms are second, nitrogen atoms are third, and chlorine and bromine fourth and fifth. Examples of molecules which contain non-polar covalent bonds are H_2 and F_2. In these cases, the connected atoms are equal in electronegativity (obviously -- they're identical atoms!), so the electrons in the bond are arranged symmetrically, and no partial electrical charges are present.

The polarity of a molecule is a vector quantity -- it has a numerical value, and it points in a certain direction. Polar covalent bonds are often represented by arrows which point to the "δ-" atom, as shown for HF: $\overset{\longmapsto}{\text{H-F}}$

POLARITY OF MOLECULES:

Molecules such as HF, which contain only one polar covalent bond, can easily be seen to be polar. When molecules have more than one polar covalent bond, we must consider the polarity of the individual bonds and the directions in which their polarity vectors point in order to be able to determine whether the molecule as a whole is polar or nonpolar. A few examples are shown below.

Problem: Are molecules of H_2O polar?

Solution: The Lewis structure of H_2O is shown at the right. The molecule is bent, due to the presence of the lone pairs on the oxygen atom. As shown in the drawing, the polarity vectors point in different directions. The sum of the two polarity vectors is the large vector shown above. (To add two vectors, draw one vector so that its "tail" is at the same point as the "head" of the other vector. The sum is the new vector which goes from the "tail" of the second vector to the "head" of the first vector.) This vector indicates that H_2O molecules have a positive end and a negative end, and are therefore polar molecules.

Problem: Are molecules of CO_2 polar?

Solution: The Lewis structure of CO_2 is shown at the right. Notice that while each C=O bond is itself polar, the two polarity vectors point in exactly opposite directions. Adding these two vectors produces a sum of zero. Therefore, molecules of CO_2 are nonpolar.

Problem: Are molecules of SO_2 polar?

Solution: The Lewis structure of SO_2 is shown at the right. (Actually, this is just one of the two possible resonance structures.) The lone pair on the sulfur atom imparts a bent geometry to the SO_2 molecule. The polarity vectors do not cancel each other out. SO_2 is polar.

VALENCE BOND THEORY:

Atoms of carbon contain a total of six electrons. The electron configuration of a carbon atom is represented by the energy-level diagram at the right. Notice that there are two unpaired electrons in the 2p subshell of a carbon atom. It would be reasonable to expect carbon atoms to form two covalent bonds by sharing those electrons with two other atoms having one unpaired electron each -- for example, hydrogen atoms. The compound formed when hydrogen and carbon react should have the formula CH_2 by the above reasoning. In fact, molecules of CH_2 are very unstable. A much more stable molecule is CH_4 (methane). Methane molecules have a tetrahedral geometry -- all of the H-C-H bond angles are 109.5^{o}. This is also difficult to explain using the above logic -- p orbitals are arranged at 90° angles to each other. Obviously, a better theory is needed to explain the bonding in CH_4 and similar compounds.

2p ' ' _
2s "
1s "

The valence bond theory was proposed by Heitler and London in 1927. It says that since electrons in atoms are located in orbitals, and since electrons are "shared" between two atoms in a covalent bond, then covalent bonds can only be made by bringing orbitals together so that they overlap. The greater the overlap of the atomic orbitals, the stronger the covalent bond. However, atomic orbitals are just the graphs of the Schrödinger wave equations. They can therefore be treated just like any other mathematical functions -- they can be added, subtracted, multiplied, etc. When these operations are performed on mathematical functions, new functions result. Similarly, when atomic orbitals from the same atom are mathematically combined, new orbitals result. These new orbitals are called hybrid orbitals. Hybrid orbitals resemble the orbitals from which they are made. For example, an s orbital and a p orbital can be "hybridized" to give two sp hybrid orbitals, which are roughly spherical (like s orbitals) but have two lobes each (like p orbitals).

HYBRID ORBITALS:

The following rules should be obeyed whenever it is necessary to combine atomic orbitals into hybrid orbitals to explain the bonding in a molecule.

Rule # 1: The orbitals being hybridized must all come from the same atom. Orbitals from different atoms are never hybridized together.

Rule # 2: Only orbitals in the valence shell of an atom may be hybridized. (This should make sense, since valence electrons form covalent bonds.)

Rule # 3: Orbitals that are relatively low in energy are hybridized before orbitals that are relatively high in energy. s orbitals and p orbitals are hybridized before d orbitals and f orbitals.

Rule # 4: Any number of orbitals may be hybridized. However, the number of hybrid orbitals obtained at the end of the hybridization process must be the same as the number of atomic orbitals used at the beginning of the process. For example, an s orbital and a p orbital (a total of two orbitals) may be combined to form a set of two "sp" hybrid orbitals.

Rule # 5: The hybrid orbitals formed by the hybridization process will arrange themselves around their atom so as to be as far apart from each other as possible. For example, the two sp hybrid orbitals formed in the example in Rule 4, above, will point in exactly opposite directions. The 180° angle between them implies a linear geometry for the covalent bonds attached to that particular atom.

The geometries of other sets of common hybrid orbitals are given below.

sp^2 hybrid orbitals: three orbitals, triangular geometry, 120° angles.

sp^3 hybrid orbitals: four orbitals, tetrahedral geometry, 109° angles.

sp^3d hybrid orbitals: five orbitals, trigonal bipyramidal, 90° & 120°.

sp^3d^2 hybrid orbitals: six orbitals, octahedral geometry, 90° angles.

(Note: "sp^2" is pronounced "S-P-two", not "S-P-squared".)

HOW TO DETERMINE THE HYBRIDIZATION OF AN ATOM:

Problem: What kind of hybrid orbitals are used by carbon in CH_4 ?

Solution: If you look at the Lewis structure at the right, you should see that the carbon atom in CH_4 is surrounded by a total of four pairs of electrons (four covalent bonds). Each pair of electrons must be located in an orbital, so a set of four orbitals is needed by the carbon atom in order to bond with four hydrogen atoms. A set of sp^3 hybrid orbitals is made from an s orbital and three p orbitals -- a total of four orbitals. Thus, the set of sp^3 hybrid orbitals must contain four orbitals, which is just the number the carbon atom needs. Therefore, the carbon atom in CH_4 is said to be sp^3-hybridized. Since sp^3 hybrid orbitals have a tetrahedral geometry, we can safely predict that the CH_4 molecule will be tetrahedral as well.

```
  H
  ..
H:C:H
  ..
  H
```

Problem: What kind of hybrid orbitals are used by chlorine in ClF_3 ?

Solution: See the Lewis structure at the right. The Cl atom is surrounded by a total of five electron pairs (two lone pairs and three covalent bonds). Therefore, a set of five hybrid orbitals is needed, which means that five atomic orbitals must have been used to make the set of hybrid orbitals. The five orbitals of lowest energy in the valence shell of a chlorine atom are the s orbital, the three p orbitals, and one of the five d orbitals. s + p + p + p + d = a set of sp^3d hybrid orbitals. Cl is sp^3d-hybridized in ClF_3.

```
 ..  ..  ..
:F:Cl:F:
 ..  ..  ..
    :F:
     ..
```

Problem: What are the hybridizations of carbon and oxygen in CO_2 ?

Solution: Multiple bonds are counted as only one pair of electrons when determining hybridizations. Therefore, the carbon atom has two pairs of electrons around it (two double bonds), so it is sp-hybridized. Each oxygen atom has three electron pairs around it (two lone pairs and the double bond), so each oxygen atom is sp^2-hybridized.

```
:O=C=O:
 ..  ..
```

SIGMA BONDS AND PI BONDS:

The reason that multiple bonds are counted as one pair of electrons (just like single bonds and lone pairs) when determining the hybridization of an atom is that multiple bonds are composed of two different types of covalent bonds. One type is made from the hybrid orbitals surrounding the atom, and the other is made from the "leftover" (unhybridized) atomic orbitals. Each multiple bond contains only one of the type of bond made from hybrid orbitals, so it is counted only as a single bond -- what we're really counting are the hybrid orbitals!

The two types of covalent bonds are sigma (lower-case Greek letter σ) bonds and pi (lower-case Greek letter π) bonds. In a sigma bond, the electrons which make up the bond are most likely to be found directly between the connected atoms. In a pi bond, the electrons which make up the bond are most likely to be found above, below, or alongside the connected atoms.

Single bonds, which are formed by the "end-to-end" overlapping of atomic orbitals, are simple examples of sigma bonds. Multiple bonds also contain sigma bonds, but they also contain one or more pi bonds. The pi bonds are formed by the "sideways" overlapping of the atomic orbitals which remain after the hybridization process is complete. Specifically, a single bond is made of one sigma bond, a double bond is made of one sigma bond and one pi bond, and a triple bond is made of one sigma bond and two pi bonds. (To help you visualize this, consider a hot dog in a bun. The hot dog looks something like a sigma bond -- the "meat" lies directly between the two ends of the hot dog. The bun looks something like a pi bond -- it lies above and below the hot dog, but not directly between the two ends of the hot dog. It is also important to note that one pi bond consists of two regions in space where electrons may be found, just as a hot dog bun consists of two pieces of bread -- the top piece and the bottom piece.)

HOW PI BONDS ARE FORMED:

The valence bond theory states that covalent bonds are formed whenever atomic orbitals overlap. The difference between sigma bonds and pi bonds is that sigma bonds are formed by the "end-to-end" overlapping of hybrid orbitals, while pi bonds are formed by the "side-to-side" overlapping of unhybridized orbitals.

Consider an atom which is sp^2-hybridized. Since the valence shells of most atoms contain one s orbital and three p orbitals, a p orbital will be "left over" after a set of three sp^2 hybrid orbitals has been made from the s orbital and two of the three p orbitals. The three sp^2 hybrid orbitals lie in the same plane, arranged in a triangular fashion (120^0 apart). The "leftover" p orbital is perpendicular to the sp^2 hybrid orbitals, projecting above and below the plane which contains them. If two sp^2-hybridized atoms are brought together so that one sp^2 hybrid orbital from one atom overlaps with one sp^2 hybrid orbital from the other atom in an "end-to-end" fashion, the two overlapping sp^2 hybrid orbitals form a sigma bond. However, a pi bond will be formed at the same time, since the two "leftover" p orbitals on the two atoms will overlap in a "side-to-side" fashion. The result is a double bond -- one sigma bond and one pi bond.

Triple bonds are formed in a similar way, using sp-hybridized atoms. The two sp hybrid orbitals are arranged in a linear fashion (180^0 apart). There are two "leftover" p orbitals -- one projects above and below the line on which the two sp hybrid orbitals lie, and the other projects in front of and behind the two sp hybrid orbitals. A sigma bond is formed when two sp-hybridized atoms are brought together so that an sp hybrid orbital on one atom overlaps with an sp orbital on the other atom. Simultaneously, the p orbitals which project above and below the sigma bond overlap to form one pi bond, and the p orbitals which project in front of and behind the sigma bond overlap to form the other pi bond.

GASES AND THEIR PROPERTIES:

Most of the matter on earth occurs in one of three physical states, or phases -- namely, the solid state, the liquid state, or the gaseous state. Gases obey some very simple physical laws, which makes it somewhat easier to study gases as a group than to study liquids and solids (the condensed phases).

When you blow up a balloon, the air inside the balloon has the same shape as the balloon does. Because gases tend to take on the same shape as their containers, they are classified as fluids. The volume of a sample of a gas is just the volume of its container. The pressure of a gas is simply the ratio of the force exerted by the gas particles (atoms or molecules) against the walls of their container to the surface area of those walls. (Pressure of any kind is defined as force/area.) Hence, gas pressure is often measured in pounds per square inch. (Atmospheric pressure is roughly 14.7 p.s.i. at sea level.)

In 1643, the Italian physicist Evangelista Torricelli showed that air exerts pressure by filling a long glass tube which had been sealed at one end with mercury, and then inverting the tube into a container of mercury which was open to the atmosphere. Some of the mercury moved down into the container, but most of it remained in the glass tube, held in place by the pressure of the air! Torricelli found that the height of the mercury column in the glass tube was approximately 760 millimeters above the surface of the mercury in the container. Hence, another common unit used to measure pressure is mmHg ("millimeters of mercury") or inHg ("inches of mercury"). "Milliliters of mercury" are sometimes called Torr in honor of Torricelli. The most common unit used by scientists to measure pressure is the atmosphere (atm). One standard atmosphere = 760 mmHg (760 Torr) = 14.7 p.s.i. = 29.92 inHg. ("Inches of mercury" are used by meteorologists to report the barometric pressure. Torricelli's apparatus is called a barometer.)

BOYLE'S LAW:

In 1666, the Irish chemist Robert Boyle performed the first quantitative experiments on gases. Boyle used a sealed glass tube similar to the one Torricelli used, but bent into the shape of the letter "J", with the sealed end at the short stem of the "J". By pouring mercury into the open end of the tube, Boyle found that he was able to "trap" a small of air in the sealed end of the tube. Furthermore, the volume of this trapped sample of air decreased when more mercury was added to the tube -- that is, when the pressure on the sample of air increased. Careful measurements of the pressure and volume of the gas sample at several different pressures and volumes led Boyle to the conclusion that volume and pressure are inversely proportional -- when one goes up by a certain amount, the other goes down by the same amount. Boyle's law can be written as:

$$PV = \text{constant}$$

where P is the pressure and V is the volume of the sample of gas. Since the product of the pressure and the volume is a constant, another way to write this is:

$$P_i V_i = P_f V_f$$

where the subscript "i" refers to the initial values of pressure and volume for the gas sample, and the subscript "f" refers to their final values.

Problem: A balloon has a volume of 2.54 L at 1.00 atm pressure. If it is taken to a higher altitude where the pressure is only 0.750 atm, what will be the volume of the balloon?

Solution: Rewriting the second equation above, we get:

$$V_f = \frac{P_i V_i}{P_f}$$

Now, just substitute in the correct values of P_i, V_i, and P_f:

$$V_f = \frac{(1.00 \text{ atm}) \times (2.54 \text{ L})}{(0.750 \text{ atm})} = 3.39 \text{ L.}$$

Note: The balloon increases in volume, as it should at the lower pressure.

CHARLES' LAW:

In 1787, the French physicist Jacques Charles measured the volumes of samples of various gases at different temperatures. He plotted his data on a graph of volume vs. temperature, and noted that the data points for each gas lay on a straight line. The lines for different gases had different slopes.

Charles drew the following conclusions from these results:

1) The volume of a sample of a gas is directly proportional to the temperature of the gas. As the temperature increases, the volume also increases. In equation form, Charles' law can be written as:

$$V = (\text{constant}) \times T$$

where V is the volume and T is the Kelvin temperature of the sample of the gas.

2) The lines on the graph all converged at the point on the graph which corresponded to a volume of zero and a temperature of -273 ^{o}C. This is the lowest possible temperature, called absolute zero (zero on the Kelvin scale).

Problem: A balloon has a volume of 750 mL in a freezer whose temperature is -10 ^{o}C. If the balloon is removed from the freezer and allowed to warm to room temperature (25 ^{o}C), what will be its volume?

Solution: A convenient form of Charles' law to use in this case is:

$$\frac{V_i}{T_i} = \frac{V_f}{T_f}$$

where "i" and "f" indicate the initial and final conditions, respectively. Using the above temperatures as written leads to a negative value for the new volume. This isn't physically possible -- there's no such thing as a "negative volume"! Converting the Celsius temperatures into Kelvin temperatures solves this problem:

$$-10 + 273 = 263 \text{ K} = T_i \qquad 25 + 273 = 298 \text{ K} = T_f$$

$$V_f = \frac{V_i T_f}{T_i} = \frac{(750 \text{ mL}) \times (298 \text{ K})}{(263 \text{ K})} = 850 \text{ mL}.$$ (Note: The balloon expands.)

THE COMBINED GAS LAW:

Boyle's law and Charles' law can be combined into one equation. This equation is sometimes called the <u>combined gas law</u>, and it is shown below:

$$\frac{PV}{T} = \text{(constant)}$$

where P is the pressure of the gas, V is the volume of the gas, and T is the <u>Kelvin</u> temperature of the gas. Another way to write this equation is:

$$\frac{P_iV_i}{T_i} = \frac{P_fV_f}{T_f}$$

where "i" and "f" refer to the <u>initial</u> and <u>final</u> conditions, respectively. Notice that if the temperature is kept constant (that is, if $T_i = T_f$), then this equation is simply the equation for Boyle's law ($P_iV_i = P_fV_f$). Similarly, if the pressure is kept constant ($P_i = P_f$), then this equation becomes the Charles' law equation.

<u>Problem</u>: A balloon has a volume of 810 mL at a pressure of 750 Torr and a temperature of -77 ^oC. (This is the temperature of a mixture of "dry ice" (frozen CO_2) and acetone.) The balloon is removed from the dry ice/acetone bath and allowed to warm to room temperature (25 ^oC) in a decompression chamber in which the pressure is 3.00 atmospheres. Calculate the final volume of the balloon.

<u>Solution</u>: This problem is difficult to solve intuitively -- warming the balloon should cause it to expand, but increasing the pressure should cause it to contract. However, this problem is easily solved using the combined gas law. We need to remember to use only <u>Kelvin</u> temperatures (no negative numbers allowed!) and to be consistent in all other units (760 Torr = 1.00 atm). Hence:

$-77 + 273 = 196\ K = T_i \qquad 25 + 273 = 298\ K = T_f$

$3.00 \text{ atmospheres} \times 760 \text{ Torr/atm} = 2280 \text{ Torr} = P_f$

$$V_f = \frac{P_iV_iT_f}{T_iP_f} = \frac{(750 \text{ Torr}) \times (810 \text{ mL}) \times (298 \text{ K})}{(196 \text{ K}) \times (2280 \text{ Torr})} = 405 \text{ mL.}$$

AVOGADRO'S LAW:

In 1808, the French chemist Joseph Gay-Lussac discovered that when gases react with each other, the volumes of the gases which react are related to each other by small whole-number ratios. For example, when hydrogen gas reacts with oxygen gas to form water vapor, two liters of hydrogen are consumed for each liter of oxygen that is consumed, and two liters of water vapor are formed in this process. The point is that the ratios by volume of hydrogen, oxygen, and water vapor involved are 2:1:2 -- small whole numbers. (This principle is sometimes called the Law of Combining Volumes.)

At about this same time, the English scientist John Dalton proposed his atomic theory, which stated that atoms combine in small whole-number ratios to form molecules during a chemical reaction. In 1811, the Italian chemist Amadeo Avogadro showed that the law of combining volumes and the atomic theory would both make sense if equal volumes of different gases contained equal numbers of particles (atoms or molecules) of the gases. For example, in the case above, two molecules of H_2 react with one molecule of O_2 to form two molecules of H_2O -- the ratio is 2:1:2, whether we're talking about volumes of gases (in liters) or amounts of gases (in numbers of molecules). The underlined principle above is sometimes called Avogadro's law, and can be written in equation form as follows:

$$V = (\text{constant}) \times n$$

where V is the volume of the gas and n is the amount of gas present, in moles (the usual units used for counting small things like molecules!). This shouldn't be too surprising -- if you want to increase the volume of a balloon, one way to do it is to blow more air into it! However, it is important to realize that the volume of the balloon depends on the amount of gas present (in moles), not on the mass of the gas or the atomic weight or molecular weight of the gas particles.

THE IDEAL GAS LAW:

The equations for Boyle's law, Charles' law, and Avogadro's law can be combined into one equation, as shown below. This equation is called the ideal gas law, and any gas which obeys this law is called an ideal gas.

Boyle: $PV = \text{constant}$ Charles: $\frac{V}{T} = \text{constant}$ Avogadro: $\frac{V}{n} = \text{constant}$

Ideal Gas Law: $\frac{PV}{nT} = \text{universal gas constant} = R = 0.082056 \frac{\text{L atm}}{\text{K mole}}$.

(It should be noted that most real gases do not obey the ideal gas law exactly, but many of them come reasonably close to ideal behavior under most conditions.)

The ideal gas law is usually written as $PV = nRT$. The other gas laws can be obtained from the ideal gas law by simply keeping all quantities constant except the two quantities being studied. For example, if pressure and temperature are kept constant, $PV = nRT$ simply becomes $V = n \times (\text{constant})$, which is Avogadro's law. The units of the gas constant (R) may look strange, but remembering them is useful, since the correct units for each quantity must be used when using $PV = nRT$.

Problem: What volume is occupied by 1.00 mole of an ideal gas at a temperature of 0 ^{o}C and a pressure of 1.00 atmosphere?

Solution: Rearranging $PV = nRT$ gives us:

$$V = \frac{nRT}{P}$$

For the units to cancel properly, the temperature must be in Kelvins:

$$T = 0 + 273 = 273 \text{ K}$$

$$V = \frac{(1.00 \text{ mole}) \times (0.082056 \frac{\text{L atm}}{\text{K mole}}) \times (273 \text{ K})}{(1.00 \text{ atm})} = 22.4 \text{ L}.$$

The temperature and pressure given above are referred to as the standard temperature and pressure, or STP for short. The volume calculated above is called the standard molar volume of an ideal gas -- that is, the volume occupied by one mole of an ideal gas at standard temperature and pressure.

USING THE IDEAL GAS LAW:

Problem: A 50.00-mL sealed glass tube contains 88.0 mg of an unknown gas at 27.0 °C and 750 Torr. What is the molecular weight of the gas?

Solution: Rearranging PV = nRT allows us to find the amount of gas present (in moles). Remembering to use the correct units gives us the following:

$$n = \frac{PV}{RT} \qquad P = 750 \text{ Torr} \times \frac{1 \text{ atm}}{760 \text{ Torr}} = 0.987 \text{ atm}$$

$$V = 50.00 \text{ mL} = 50.00 \times 10^{-3} \text{ L}$$

$$T = 27.0 + 273 = 300 \text{ K}$$

$$n = \frac{(0.987 \text{ atm}) \times (50.00 \times 10^{-3} \text{ L})}{(0.082056 \frac{\text{L atm}}{\text{K mole}}) \times (300 \text{ K})} = 2.00 \times 10^{-3} \text{ moles}$$

Molecular weight has units of g/mole, so divide the mass of the gas by this value:

$$88.0 \text{ mg} = 88.0 \times 10^{-3} \text{ grams}$$

$$\text{Molecular Weight} = \frac{88.0 \times 10^{-3} \text{ grams}}{2.00 \times 10^{-3} \text{ moles}} = 44.0 \text{ g/mole.}$$

This is an example of the Dumas method for determining the molecular weight of a gas. One possible identity for the gas is CO_2 (12 + 16 + 16 = 44).

Problem: Calcium carbonate decomposes upon heating. The reaction is:

$$CaCO_{3\,(s)} \longrightarrow CaO_{(s)} + CO_{2\,(g)}$$

What volume of CO_2 at STP is produced when 304 grams of $CaCO_3$ is decomposed?

Solution: The molecular weight of $CaCO_3$ is 100.09 g/mole. Therefore:

$$(304 \text{ g } CaCO_3) \times \frac{1 \text{ mole } CaCO_3}{100.09 \text{ g } CaCO_3} \times \frac{1 \text{ mole } CO_2}{1 \text{ mole } CaCO_3} = 3.04 \text{ moles } CO_2$$

The ideal gas law could be used at this point, but since we're working at STP, we can simply use the standard molar volume of a gas: 22.4 L = 1 mole of gas at STP.

$$3.04 \text{ moles } CO_2 \times \frac{22.4 \text{ L}}{1 \text{ mole}} = 68.0 \text{ L.}$$

Note: This "trick" only works at standard temperature and pressure! At other temperatures and pressures, the ideal gas law must be used.

DALTON'S LAW OF PARTIAL PRESSURES:

In the early part of the 19th century, the English scientist John Dalton (the same John Dalton who proposed the atomic theory!) was studying mixtures of gases. Not surprisingly, he found that each gas in a mixture of gases expands to fill its container. (This should make sense -- consider the air in the room where you are now. Air is a mixture of nitrogen, oxygen, and other gases, all of which expand to fill the room. Anywhere you go in the room, you can find oxygen to breathe -- fortunately!) Therefore, the pressure exerted by one gas in a mixture of gases is the same as it would be if the gas were the only gas present in the container. Dalton called the pressures of the individual gases in the gas mixture partial pressures. Furthermore, he determined that the total pressure exerted by a mixture of gases is equal to the sum of the partial pressures of the individual gases in the mixture. In equation form, this can be written as:

$$P_m = P_1 + P_2 + P_3 + P_4 + \dots$$

where P_m is the total pressure of the mixture of gases and P_1, P_2, P_3, P_4, and so on are the partial pressures exerted by gas # 1, gas # 2, gas # 3, gas # 4, etc.

One way to prepare samples of pure gases in the laboratory is to do a chemical reaction which produces the desired gas and pass the gas from the vessel in which the reaction is occurring into a bottle filled with water. The gas displaces the water, leaving the pure gas in the bottle. However, this gas isn't really "pure" -- it's mixed with water vapor! As a liquid evaporates, the vapors that are produced behave like any other gas. For instance, the molecules of the vapors collide with the walls of their container, exerting a pressure which is called the vapor pressure of the liquid from which they came. As the temperature of a liquid rises, its vapor pressure rises. The liquid's boiling point is the temperature at which its vapor pressure equals the atmospheric pressure.

GRAHAM'S LAW OF EFFUSION:

If you've ever owned a helium-filled balloon, you've probably noticed that helium-filled balloons deflate more rapidly than air-filled balloons do. In either case, the balloon deflates because the gas particles (atoms or molecules) inside the balloon gradually escape from the balloon through small pores in the walls of the balloon. This process is called <u>effusion</u>. Effusion techniques were used during World War II to separate the isotopes of uranium.

In 1829, the Scottish chemist Thomas Graham discovered that the rate at which a gas undergoes effusion is <u>inversely</u> proportional to the <u>square root</u> of the <u>molecular weight</u> of the gas. In equation form, this can be written as:

$$\text{Rate of Effusion} = \frac{\text{constant}}{(\text{M.W. of gas})^{1/2}}$$

This can also be written as follows when comparing the effusion rates of two gases:

$$\frac{\text{Rate of Effusion, Gas \# 1}}{\text{Rate of Effusion, Gas \# 2}} = \frac{(\text{M.W. of Gas \# 2})^{1/2}}{(\text{M.W. of Gas \# 1})^{1/2}}$$

<u>Problem</u>: Which undergoes effusion at a faster rate -- He or CH_4 ? Calculate the relative rates of effusion of these two gases.

<u>Solution</u>: Use the above equation. Let's make He = Gas # 1:

$$\frac{\text{Rate of Effusion of He}}{\text{Rate of Effusion of } CH_4} = \frac{(\text{M.W. of } CH_4)^{1/2}}{(\text{M.W. of He})^{1/2}} = \frac{(16.043 \text{ g/mole})^{1/2}}{(4.00260 \text{ g/mole})^{1/2}}$$

$$= 4.0054/2.0006 = 2.0020$$

Thus, it can be seen that helium undergoes effusion about twice as rapidly as CH_4 does. This should make sense -- helium atoms are lighter (less massive) than CH_4 molecules. Therefore, if a helium atom and a methane molecule have the same amount of kinetic energy (K.E. = $mv^2/2$), the less massive particle (helium) should have the greater velocity. The relative rates of effusion of two gases are determined by the relative velocities with which their particles move.

THE KINETIC-MOLECULAR THEORY OF GASES:

In the 1860's, the Scottish physicist James Clerk Maxwell and the German physicist Ludwig Boltzmann developed a model to try to explain why gases behave as they do. This model is now called the kinetic-molecular theory of gases. The essential ideas of the kinetic-molecular theory are outlined below:

1. Ideal gases consist of particles which are so small (compared with the relatively large distances between them) that they can be considered to have essentially zero volume. Gases have volumes because of the relatively large distances between the gas particles, not because of the size of the particles.

2. The particles of an ideal gas do not attract or repel each other. However, they do move around constantly, and occasionally may collide with each other and with the walls of their container. Gases have pressures due to the force exerted by the collisions of their particles with the walls of their containers.

3. During a collision between two particles, kinetic energy may be transferred from one particle to another. Hence, two different particles of the same sample of a gas may have different kinetic energies. The average kinetic energy of a sample of gas particles is directly proportional to the temperature of the gas, in Kelvins.

This model can be used to explain many of the properties of gases. For example, gases are compressible, but liquids and solids are not compressible. This is because compressing a gas simply forces the particles of the gas closer together. The particles in liquids and solids are already close together, so that they cannot be compressed. Also, why should there be a "coldest possible temperature" (absolute zero) ? The answer is that as the temperature decreases, the kinetic energy of the gas molecules decreases, causing their velocity to decrease. Absolute zero is the temperature at which the molecules stop moving!

THE KINETIC-MOLECULAR THEORY AND THE FUNDAMENTAL GAS LAWS:

The principles of the kinetic-molecular theory of gases can be used to explain many of the fundamental gas laws. Examples are given below.

Boyle's Law: The pressure exerted by a gas comes from the force with which the gas particles strike the walls of their container. If the volume of the container increases, the surface area of the inner walls of the container increases as well. If the number of particles present and the temperature remain constant, the total force exerted against the walls of the container remains the same, so the pressure (the average force per unit of surface area) decreases.

Charles' Law: As the temperature of a gas increases, the kinetic energy of the gas particles increases. Since K.E. = $mv^2/2$, and since the mass (m) of the gas particles doesn't change, the result is that the velocity (v) of the gas particles increases. When the rapidly-moving gas particles collide with the walls of the container, they exert a greater force on the walls than slowly-moving gas particles do. Hence, the pressure of the gas sample will increase. However, if the walls of the container are movable (for example, the flexible rubber in a balloon), the increased force pushes them out, so the volume of the gas increases.

Graham's Law: The rate of effusion of a gas is directly proportional to the velocity of the gas particles. Since K.E. = $mv^2/2$, the velocity of the gas particles can be expressed as: $v = [(2) \times (K.E.)/m]^{1/2}$. If the temperature of the gas remains constant, the value of K.E. will remain constant. Therefore, the rate of effusion of the gas will be directly proportional to $m^{-1/2}$, where m is the mass of the gas particles -- that is, the molecular weight of the gas!

Similar arguments can be made for Avogadro's Law and for Dalton's Law. This is another illustration of the principal work of the scientist -- to construct a model which accurately predicts and/or explains the observations made in nature!

LIQUIDS AND SOLIDS:

The behavior of gases can be described by relatively simple models such as the kinetic-molecular theory because the individual gas particles act independently of each other. The attractive forces between the gas particles are relatively small, due to the relatively large distances between them. However, in the condensed phases (liquids and solids), the distances between the particles are much smaller, and the attractive forces between the particles are therefore much greater. Because of this, the particles do not act independently of each other. Therefore, no simple model will accurately describe the behavior of liquids and solids. However, some generalizations can be made about their behavior.

Unlike gases, liquids and solids cannot be compressed. (An important feature of the braking system of most cars is the transmission of force from the brake pedal to the brake shoes through the brake fluid (a liquid). If liquids were compressible, this force could not be transmitted, and the brakes wouldn't work.) Unlike solids, liquids are fluids -- they take on the shapes of their containers, and they have flow rates which can be measured. (Gases are also fluids.) A unique feature of liquids is their surface tension, which is caused by the attraction of liquid particles below the liquid surface for liquid particles at the surface.

A phase diagram is a graph of pressure vs. temperature which summarizes the conditions under which a substance exists as a solid, as a liquid, or as a gas. Not surprisingly, most substances are solids at high pressures and low temperatures and gases at high temperatures and low pressures, with liquids being favored at intermediate conditions of pressure and temperature. At the boundary lines between the solid, liquid, and gas regions of a phase diagram, more than one phase can exist. At the point where all three regions meet on the phase diagram (called the triple point), all three phases exist simultaneously.

PHASE TRANSITIONS AND DYNAMIC EQUILIBRIUM:

When a substance undergoes a change from one physical state to another, the process is referred to as a phase transition. Common examples of phase transitions include the melting of ice (solid to liquid), the freezing of water (liquid to solid), the boiling of water (liquid to gas), and the condensation of water to form dew (gas to liquid). The process in which a solid is converted into a gas directly (without passing through a liquid phase) is called sublimation. (Solids which have odors, such as solid air fresheners, sublime to a small degree.)

Phase transitions depend upon the kinetic energies of the particles involved and the attractive forces between the particles. For example, evaporation occurs when liquid particles with relatively high kinetic energy approach the liquid surface and enter the gas phase by overcoming the attractive forces in the liquid phase. Similarly, condensation occurs when gas particles strike a liquid or solid surface and transfer some of their kinetic energy to that surface. The particles no longer have enough kinetic energy to overcome the attractive forces in the condensed phase, so they remain in the liquid or solid phase.

Now, consider a butane cigarette lighter. Butane (C_4H_{10}, b.p. 0 ^oC) is normally a gas at room temperature, but the lighter contains both liquid butane and gaseous butane in contact with each other. Butane molecules continually move back and forth between the liquid phase and the gas phase. Thus, both vaporization and condensation (two opposing phase transitions) are occurring simultaneously. The system as a whole does not change, but individual molecules are constantly changing phases. This situation is one example of a dynamic equilibrium, and is written as:

$$C_4H_{10}\ (l) \rightleftarrows C_4H_{10}\ (g)$$

The arrows pointing in each direction indicate that two opposing phase transitions are occurring simultaneously. The system is dynamic (changing) but at equilibrium.

LeCHATELIER'S PRINCIPLE:

Consider the following equilibrium, which occurs in a butane lighter:

$$C_4H_{10}\,(l) \rightleftharpoons C_4H_{10}\,(g)$$

When the lighter is ignited, it is the butane gas (nearest the top of the lighter) which reacts with oxygen in the air to produce the lighter flame. Therefore, the amount of gaseous butane in the lighter decreases. This decreases the likelihood that a molecule of butane in the gas phase will strike the liquid butane surface and re-enter the liquid phase. Therefore, the two phase transitions above (liquid to gas and gas to liquid) are no longer equally probable, and the system is not at equilibrium any more. However, the system can easily re-establish the equilibrium if some of the liquid butane evaporates to provide a fresh supply of butane gas. In fact, this is what happens -- the lighter works until the liquid butane is gone.

This is a specific example of a more general principle that applies to systems in a state of dynamic equilibrium. Dynamic equilibrium is a low-energy state, and any system will try to minimize its energy whenever possible. Thus, if a system in a state of dynamic equilibrium is disturbed in such a way so that it is no longer at equilibrium, then the system will readjust itself (if possible) to try to counteract the disturbance and restore the equilibrium. This principle was first stated in 1888 by the French physical chemist Henri Louis LeChatelier, and is known as LeChatelier's Principle.

In the above case, the "disturbance" is the ignition of the lighter. This consumes $C_4H_{10}\,(g)$ and removes it from the system. "Disturbances" in the sense of LeChatelier's Principle generally change the amount of material on one side of the equation. The "readjustment" is the liquid-to-gas phase transition of butane, restoring the gaseous butane which was removed by the "disturbance". When the liquid butane runs out, this "adjustment" is no longer possible.

HYDROGEN BONDING:

If the intermolecular attractive forces which hold liquid molecules in the liquid phase are relatively <u>strong</u>, then a greater amount of kinetic energy will be needed for an individual molecule to overcome these forces and enter the gas phase. The greater the <u>temperature</u> of the molecules, the greater their kinetic energy. At high temperatures, the vapor pressure of a liquid is greater than at low temperatures, since at high temperatures, more liquid molecules have enough kinetic energy to overcome the intermolecular attractive forces and enter the gas phase. Therefore, a compound's <u>boiling point</u> (the temperature at which its vapor pressure equals 1.00 atmosphere) can be useful in determining the strength of the intermolecular attractive forces which hold its molecules in the liquid phase.

Consider water (H_2O, b.p. 100 ^oC) and hydrogen sulfide (H_2S, b.p. -60 ^oC). Since higher temperatures are needed to cause water to boil than to cause hydrogen sulfide to boil, we can conclude that the intermolecular attractive forces between water molecules are stronger than the intermolecular attractive forces between hydrogen sulfide molecules. This is due to the fact that water molecules are much more <u>polar</u> than hydrogen sulfide molecules. Oxygen atoms are much more electronegative than sulfur atoms, and will therefore carry a stronger partial negative charge than sulfur atoms. This means that the hydrogen atoms in water molecules will carry a stronger partial positive charge than the hydrogen atoms in hydrogen sulfide molecules. The strong partial positive charges on the hydrogen atoms in water molecules will be attracted to the strong partial negative charges on the oxygen atoms in other water molecules. This causes water molecules to be more attracted to each other than hydrogen sulfide molecules are. The attraction of a partially positive hydrogen atom for a partially negative atom in another molecule is called <u>hydrogen bonding</u> (a relatively <u>strong</u> intermolecular force).

INTERMOLECULAR ATTRACTIVE FORCES:

Compare the boiling points of the compounds below:

```
                                                         HHH
                                                         """
H H H H           H H H H           H H H H          H C H
" " " "           " " " "           " " " "          " " "
:O:C:C:O:         :O:C:C:C:H        H:C:C:C:C:H      H:C:C:C:H
   " "               " " "            " " " "          " " "
   H H               H H H            H H H H          H H H
Ethylene Glycol   Propanol          Butane           Isobutane
b.p. 198 °C       b.p. 97 °C        b.p. 0 °C        b.p. -12 °C
```

Notice that the boiling points decrease as the amount of hydrogen bonding (that is, the number of O-H bonds present) decreases. (Ethylene glycol (anti-freeze) has two O-H bonds, propanol has one O-H bond, and butane and isobutane have no O-H bonds.) Hydrogen bonding is one example of a class of intermolecular attractive forces called dipole-dipole interactions. Dipole-dipole interactions occur in molecules that have polar bonds, and consist simply of the attraction between a partially positive atom in one molecule and a partially negative atom in another molecule. This is not restricted to bonds involving hydrogen atoms; for example, molecules of CO_2 are held together in the solid state ("dry ice" is solid CO_2) by means of dipole-dipole interactions. (CO_2 is a nonpolar molecule but contains polar bonds.)

London forces are the primary intermolecular attractive forces between molecules which do not contain polar bonds. These forces come about because the electrons in atoms and molecules are constantly in motion. When a majority of the electrons in a molecule move to one side of the molecule, that side of the molecule takes on a partial negative charge, leaving a partial positive charge on the other side of the molecule. These partial charges make up an instantaneous dipole -- it only exists for an instant before the electrons move again. London forces are just the attractions between instantaneous dipoles in different molecules. Elongated molecules (such as butane) have a larger surface area than more "spherical" molecules (such as isobutane), and thus have greater attractive London forces.

CRYSTALLINE SOLIDS:

Many solids are made of particles which are arranged in highly ordered, repeating patterns called lattices. Lattices are usually represented by their unit cells, which are small segments of the lattice in question. Repeating the pattern of the unit cell many times in all directions generates the entire lattice. The simplest types of lattices are those based on cubic unit cells. The simplest cubic unit cell is simply eight particles arranged at the corners of a cube. This is called a simple cubic unit cell. Variations on this theme include the arrangement which includes a particle at the center of the cube (body-centered cubic), and the arrangement which includes a particle at the center of each of the six faces of the cube (face-centered cubic). Other unit cells are variations of the simple cubic structures in which one side of the "cube" is longer than the others or shorter than the others, or in which some of the angles between the sides of the "cube" are greater than 90^0 or less than 90^0.

Solids whose particles are arranged in lattices are called crystalline solids (or just crystals). Crystalline solids can usually be classified as one of four basic types. Ionic crystals (such as NaCl) are composed of cations and anions which are held in their lattice sites by ionic bonds. They are hard and brittle, with high melting points. They don't conduct electricity well when solid, but the solutions made by dissolving them in water conduct electricity. Covalent crystals (such as diamonds) are composed of atoms held together by covalent bonds. They are hard crystals with high melting points, and generally don't conduct electricity. Molecular crystals (such as ice) are composed of molecules held together by dipole-dipole attractions or London forces. They are soft crystals with low melting points. Metallic crystals (metals, such as iron) are composed of cations held in place by a "sea" of electrons around them. Metals conduct electricity very well.

PROPERTIES OF CRYSTALLINE AND AMORPHOUS SOLIDS:

In an ionic crystal, the cations and anions are next to each other. This represents a very stable arrangement for these ions, since positive charges and negative charges attract each other. (This is called Coulomb's Law.) Many of the properties of ionic crystals can be explained on this basis. For example, ionic crystals generally have high melting points. (NaCl melts at over 800 °C.) This is due to the fact that the attractive forces between ions are very strong, so a large amount of energy (that is, heat) must be supplied to overcome these forces and melt the crystal. The brittleness of ionic crystals is due to the fact that ions of similar charge are forced together when the crystal is subjected to extreme pressure. These ions tend to repel each other, causing the crystal to break apart. Also, electric currents flow when charged particles (such as ions or electrons) are free to move. Ions trapped in a rigid crystalline lattice are not free to move, but become free when the crystal is melted or dissolved in water. Thus, solutions of ions conduct electricity, but ionic crystals do not.

The melting of a crystal occurs at a constant temperature if the pressure remains constant. (For example, ice melts at 0 °C at 1.00 atmosphere.) This is called a first-order phase transition. However, some substances (such as glass and plastics) do not undergo first-order (constant temperature) phase transitions, but instead gradually soften as they are heated. Solids which behave in this way are called amorphous solids. ("Amorphous" means "without shape".) The molecules of amorphous solids are usually long, "spaghetti"-like strands of atoms which do not fit conveniently into a lattice arrangement. Hence, amorphous solids are sometimes called non-crystalline solids. Since "crystallization" never really occurs for an amorphous solid, it is sometimes called a supercooled liquid. (The supercooling process cools a liquid below its freezing point without its freezing.)

MISCIBILITY OF LIQUIDS:

As the saying goes, "oil and water don't mix." This is fortunate for the makers of salad dressing! The makers of alcoholic beverages, on the other hand, take advantage of the fact that ethyl alcohol and water do mix together. The tendency of some liquids to "mix" -- that is, to form homogeneous mixtures -- is referred to as their miscibility. Two liquids are miscible if they are each soluble in the other, regardless of the relative proportions used.

Why are some liquids miscible and other liquids are not miscible? The answer has to do with the intermolecular forces of attraction between molecules of the liquids in question. For two liquids to be miscible, their intermolecular attractive forces must be of approximately equal strength. Any attempt to mix two liquids whose intermolecular attractive forces are of unequal strengths will result in the molecules of the liquid with the stronger intermolecular attractive forces tending to form relatively strong bonds to each other, leaving the molecules of the liquid with the weaker intermolecular attractive forces to form a separate layer. In short, no mixing takes place.

A simple rule which summarizes all of this is the "like dissolves like" rule, which says that liquids of similar polarity are miscible with each other, while liquids with different polarities are not miscible. For example, water (H_2O) is a polar molecule -- water molecules tend to connect to each other by forming hydrogen bonds. Liquids such as ethyl alcohol (CH_3CH_2-O-H), which also tend to form hydrogen bonds, are generally miscible with water. However, liquids which do not tend to form hydrogen bonds (such as the "oil" in salad dressing) are usually immiscible with water. (The "oil" in salad dressing is composed largely of carbon and hydrogen, which are similar in electronegativity. Thus, "oil" is nonpolar, and its molecules are attracted to each other by relatively weak London forces.)

SOLUBILITY OF GASES IN LIQUIDS:

Gases are soluble in liquids to some extent, as a moment's thought will make obvious. For example, oxygen (O_2, a gas) is transported to the brain by the blood (a liquid). The solubility of a gas in a liquid depends mostly on two properties of the gas in question -- its pressure and its temperature.

Not surprisingly, the solubility of a gas in a liquid increases as the pressure of the gas increases. Again, this should be obvious from everyday experience. For example, when you open a can of a "carbonated" beverage, you see bubbles come to the surface of the liquid. The bubbles are made of carbon dioxide (CO_2, a gas) which was dissolved in the liquid until you opened the can. Cans of carbonated beverages are sealed so as to contain a slightly higher pressure of gas than atmospheric pressure. Exposing the carbonated beverage to a lower pressure (atmospheric pressure) causes the dissolved gas (CO_2) to be less soluble in the liquid than it was, with the result that bubbles form.

Somewhat surprisingly, the solubility of a gas in a liquid decreases as the temperature of the liquid increases. (For solids, the opposite is true -- for example, sugar (a solid) is more soluble in hot water than in cold water.) Once again, you've probably seen this phenomenon before. For example, if you heat a pan of water on a stove, you will notice that just before the water boils, some relatively small bubbles will form along the insides of the walls of the pan. These bubbles are not made of water vapor (like the "boiling bubbles" are), but are actually made of oxygen gas which had been dissolved in the water before you heated it. This phenomenon is the basis for the environmental problem known as thermal pollution -- if the temperatures of the lakes and ponds where fish live rise above certain levels, the amount of oxygen dissolved in the water will be reduced below the point at which the water can support fish and other marine life.

SOLUBILITY OF SOLIDS IN WATER:

When a crystalline solid dissolves in water, it does so because water molecules strike the crystal lattice with enough force to dislodge particles of the solid, whether these particles are <u>molecules</u> (such as the $C_{12}H_{22}O_{11}$ molecules of which sugar is made) or <u>ions</u> (such as the Na^+ and Cl^- ions of which table salt -- sodium chloride -- is made). The water molecules then surround the particles of the solute, forming a "solvent cage" around them which is held in place by ion-dipole or dipole-dipole attractive forces. These "solvent cages" keep the solute particles separated from each other, preventing recrystallization of the solid.

Not all ionic compounds are soluble in water. The reasons for this are not well understood, but a series of empirical rules for predicting whether an ionic compound is soluble in water has been developed. The <u>solubility rules</u> given below are numbered 1 through 6 for future reference.

The following ionic compounds are <u>soluble in water</u>:

1. All salts containing Na^+, K^+, or NH_4^+ ions.
2. All salts containing NO_3^-, ClO_4^-, ClO_3^-, or $C_2H_3O_2^-$ ions.
3. All salts containing halide ions (Cl^-, Br^-, I^-) <u>except</u>:
   Ag^+, Hg_2^{2+}, and Pb^{2+} halides.
4. All salts containing sulfate (SO_4^{2-}) ions, <u>except</u>:
   $CaSO_4$, $BaSO_4$, $RaSO_4$, $SrSO_4$, $PbSO_4$, and Hg_2SO_4.

The following ionic compounds are <u>insoluble in water</u>, but are salts of weak acids and are therefore <u>soluble in acid</u>:

5. All salts containing oxide (O^{2-}) and hydroxide (OH^-) ions, <u>except</u>:
   Ca^{2+}, Ba^{2+}, Na^+, K^+, and NH_4^+ oxides and hydroxides.
6. All salts containing CO_3^{2-}, PO_4^{3-}, S^{2-}, or SO_3^{2-} ions, <u>except</u>
   those which are soluble according to Rule # 1 above.

ELECTROLYTES:

As you probably already know, it's not a good idea to go swimming in the ocean during a thunderstorm! Neither is it a good idea to drop an electrical appliance into your bathtub while you're bathing! This is because both the ocean and your bath water conduct electricity, which could harm you if you're careless.

Pure water does not conduct electricity, but water which contains ions does conduct electricity. Any substance which produces ions when dissolved in water, thereby creating a solution which conducts electricity, is called an electrolyte. All water-soluble ionic compounds (such as NaCl in salty ocean water) are electrolytes, but some water-soluble covalent compounds are also electrolytes. An example is HCl, which is normally a gas, but reacts with water to form ions:

$$HCl_{(g)} + H_2O_{(l)} \longrightarrow H_3O^+_{(aq)} + Cl^-_{(aq)}$$

Some covalent compounds are nonelectrolytes -- their aqueous solutions do not conduct electricity. A common example of a nonelectrolyte is table sugar (sucrose, $C_{12}H_{22}O_{11}$), whose aqueous solutions contain sucrose molecules, but not many ions.

Compounds such as NaCl and HCl are examples of strong electrolytes -- their aqueous solutions are good conductors of electricity. Some covalent compounds which are electrolytes produce aqueous solutions which are relatively poor conductors of electricity. These compounds are called weak electrolytes. Examples include ammonia (NH_3) and acetic acid ($HC_2H_3O_2$, found in vinegar). The reactions these compounds undergo when dissolved in water are shown below.

$$NH_{3\,(g)} + H_2O_{(l)} \longrightarrow NH_4^+{}_{(aq)} + OH^-_{(aq)}$$

$$HC_2H_3O_{2\,(aq)} + H_2O_{(l)} \longrightarrow H_3O^+_{(aq)} + C_2H_3O_2^-{}_{(aq)}$$

Compounds such as ammonia and acetic acid are weak electrolytes because the reactions shown above do not go to completion -- a relatively small number of ions is formed, resulting in a solution which is a relatively poor conductor.

COLLIGATIVE PROPERTIES:

In the winter, when snow and ice make the highways slippery and thus dangerous for travel purposes, rock salt is spread out over the roads in order to "melt" the ice. Actually, the salt doesn't "melt" the ice -- that is, it doesn't supply any additional heat. What happens is that the rock salt dissolves in the ice, forming a solution whose freezing point (or melting point, if you like) is lower than the freezing point of pure water. Since colder temperatures are needed for freezing to occur, the ice-salt mixture melts more readily than pure ice does.

The lowering of the freezing point of an aqueous solution of a salt is one example of several properties of solutions referred to collectively as colligative properties. What distinguishes a colligative property from any of the other properties of solutions is that colligative properties depend only on the numbers of particles of solute and solvent present in the solution, and not on the identities of the solute and the solvent. For example, the freezing point of any aqueous solution will be lower than the freezing point of pure water -- it doesn't matter whether the solute is NaCl, $CaCl_2$, $C_{12}H_{22}O_{11}$, or any other solid. However, the identity of the solute does help determine the extent to which the freezing point of the solution is lowered.

The reason for the lowering of the freezing point of a solution is that the presence of the solute particles makes it more difficult for the particles of the solvent to arrange themselves into the crystalline lattice formation needed for freezing to occur. More energy must be removed from these solvent particles -- that is, lower temperatures must be achieved -- before crystallization (freezing) can take place. Similarly, the boiling point of a solution is higher than the boiling point of the pure solvent, since the solute particles occupy some of the surface area of the solution, making vaporization of solvent less likely to occur.

OSMOTIC PRESSURE:

All living things are made of cells, which contain water in addition to various other substances. Water molecules are the smallest molecules present in living cells -- in fact, they are so small that they are able to pass through the "pores" in the membrane which surrounds the cell. Cell membranes are sometimes called semipermeable membranes for this reason -- they allow some molecules (such as water molecules) to pass through, but not others. The flow of water molecules (or molecules of any other solvent) through a semipermeable membrane is called osmosis. Generally, osmosis tends to occur in such a way as to equalize the concentrations of the solutions on opposite sides of the membrane in question -- that is, water tends to flow from a less concentrated solution into a more concentrated solution if the two solutions are separated by a porous membrane. This process can be stopped or reversed by increasing the pressure inside the more concentrated solution. (One way to obtain pure water from salt water makes use of this principle. This process is called reverse osmosis.) The pressure required to exactly stop osmosis from a pure solvent into a solution is called the osmotic pressure of that solution. The osmotic pressure of a solution can be calculated using the equation: $\Pi = MRT$, where Π (capital Greek letter pi) is the osmotic pressure, M is the molar concentration of the solution, R is the ideal gas constant ($0.082056 \frac{\text{L atm}}{\text{K mole}}$), and T is the Kelvin temperature.

Problem: What is the osmotic pressure of a 1.00 M solution at 25 ^{o}C ?

Solution: T = 25 + 273 = 298 K. Now, just use the above equation:

$$\Pi = MRT = (1.00 \frac{\text{mole}}{\text{L}}) \times (0.082056 \frac{\text{L atm}}{\text{K mole}}) \times (298 \text{ K}) = \underline{24.5 \text{ atm}}.$$

(Note: It can be shown that the osmotic pressure above is sufficient to support a column of water over 800 feet tall! This helps to explain how water is able to flow to the leaves at the tops of very tall trees.)

USING COLLIGATIVE PROPERTIES TO DETERMINE MOLECULAR WEIGHTS:

<u>Problem</u>: A solution was prepared by dissolving 5.00 g of an unknown protein in 50.00 mL of water. The resulting solution had an osmotic pressure of 1.00 Torr at a temperature of 27 ^oC. What is the molecular weight of the protein?

<u>Solution</u>: Using $\Pi = MRT$, find the concentration of the solution:

$M = \Pi/RT = (1.00 \text{ Torr}) \times (1 \text{ atm}/760 \text{ Torr}) \div [(0.082056 \frac{\text{L atm}}{\text{K mole}}) \times (27 + 273 \text{ K})]$

$= \underline{5.35 \times 10^{-5} \text{ moles/liter}}$. Multiply this by the volume of the solution:

Amount of protein = $(5.35 \times 10^{-5} \text{ M}) \times (50.00 \text{ mL}) \times (1 \text{ L}/1000 \text{ mL}) = \underline{2.67 \times 10^{-6} \text{ mol}}$.

Molecular weight has units of grams/mole, so simply divide mass by amount:

Molecular Weight of Protein = $(5.00 \text{ g}) \div (2.67 \times 10^{-6} \text{ moles}) = \underline{1.87 \times 10^{6} \text{ g/mole}}$.

(This is a typical molecular weight for a protein -- these are <u>big</u> molecules!)

<u>Problem</u>: A solution was prepared by dissolving 3.55 g of an unknown non-electrolyte in 7.7 mL of methanol (CH_4O, M.W. = 32.04 g/mol, density = 0.79). The vapor pressure of the resulting solution was 95 Torr. Pure methanol at the same temperature has a vapor pressure of 100 Torr. What is the molecular weight of the unknown compound, assuming it is not volatile?

<u>Solution</u>: Start by rewriting Raoult's Law ($P_{solution} = X_{solvt.} P_{solvt.}$):

$X_{solvent} = P_{solution}/P_{solvent} = 95 \text{ Torr}/100 \text{ Torr} = \underline{0.95}$.

Next, calculate the amount of methanol present in the solution:

Amount of Methanol = $(7.7 \text{ mL}) \times (0.79 \text{ g/mL}) \div (32.04 \text{ g/mole}) = \underline{0.19 \text{ moles methanol}}$.

The amount of the unknown compound present can be found by using the definition of mole fraction, substituting the numbers above, and solving:

$$X_{solvent} = \frac{\text{moles solvent}}{\text{total moles}} = \frac{\text{moles methanol}}{\text{moles methanol + moles compound}} = \frac{0.19}{0.19 + \text{m.c.}} = 0.95.$$

Amount of Compound = $(0.19/0.95) - 0.19 = \underline{0.01 \text{ moles of compound}}$.

The last step is similar to the last step in the first problem:

Molecular Weight of Compound = $3.55 \text{ g}/0.01 \text{ mole} = \underline{4 \times 10^{2} \text{ g/mole}}$. (<u>One</u> sig. fig.!)

COLLOIDS:

When you shine a beam of light from a flashlight through the smoke from a campfire, you can follow the path of the flashlight beam with your eyes. This is also possible with the beam of light from the headlights of a car on a foggy night. Normally, however, beams of light passing through the air are invisible. They are made visible in the above cases by the presence of particles (such as the ashes in smoke or the water droplets in fog) which are large enough to reflect the beam of light back to our eyes. (Molecules of air are too small to reflect light.) This phenomenon -- the reflection of light by small, finely-divided particles of one substance dispersed through another substance -- is called the Tyndall effect.

Mixtures which exhibit the Tyndall effect are called colloids. The major difference between colloids and solutions is the size of the particles. True solutions are made up of molecules, ions, and/or particles of similar size. Larger particles, such as the water droplets in fog, are present in colloids. However, colloids cannot be separated by gravity or by filtration, whereas suspensions can. (Examples of suspensions include muddy water or "oil-and-water" salad dressings.)

Smoke and fog are examples of a particular type of colloid called an aerosol. Aerosols are colloids in which one of the two substances present is a gas (air, in the cases above). If both substances present in the colloid are liquids, the colloid is called an emulsion. Milk is an emulsion of butterfat in water. Emulsions do not separate into layers due to the presence of emulsifiers, which are substances that are partially soluble in each substance present in the emulsion. Mayonnaise is an emulsion of oil in water, with egg yolks serving as the emulsifier.[1] Soap is also an emulsifier -- it is partially soluble in both grease and water, and therefore makes grease more water-soluble.

---

1) Cobb, Vicki, Science Experiments You Can Eat, 1972 by J. B. Lippincott Company, New York, p. 36.

CHEMICAL KINETICS:

Some reactions take place at a faster rate than others. For example, the burning of coal and the rusting of iron are basically the same reaction -- oxidation -- taking place at different speeds. The branch of chemistry that deals with the rates at which chemical reactions occur is called chemical kinetics. (It should be noted that while chemical thermodynamics can be used to predict whether or not a reaction can occur, it says nothing about how rapidly the reaction occurs. Some reactions which are spontaneous are too slow to have any practical use, so the study of chemical kinetics is important for "real-world" chemical processes.)

Many factors influence the rates at which reactions occur. Some things are just more reactive than others -- for example, carbon (coal) reacts faster with oxygen than iron does. However, reactions don't occur at all until the reactants are brought into contact with each other. The greater the extent to which the reactants are mixed, the faster the reaction. (For example, homogeneous reactions usually occur more rapidly than heterogeneous reactions, since the reactants in a heterogeneous reaction come together only at the interface of the two phases in the reaction mixture.) Reactions occur when molecules collide, so the more molecules of reactants present per unit volume in a reaction mixture, the greater the chance for reactive collisions to occur. Thus, increasing the concentrations of the reacting species generally increases the rate of the reaction. However, some of the many collisions between molecules do not lead to a reaction occurring, due to insufficient force of collision between the reacting molecules or to an incorrect orientation of the colliding molecules. Increasing the temperature of a reaction provides more energy for the colliding molecules and increases the probability of collisions occurring with the proper orientation. Lastly, catalysts are substances which, when added to a reaction, affect the reaction's rate without being consumed.

RATE EQUATIONS:

The rate at which a chemical reaction proceeds can be determined by measuring either the rate at which the products of the reaction are formed or the rate at which the reactants are consumed. In either case, the measurement of the concentrations of various substances present in the reaction mixture is done at regular intervals. The rate of the reaction is simply the rate at which the concentration (of reactants or products) changes as a function of time.

As the reaction proceeds, the concentration of the reactants decreases as the reactants are consumed. As a result, the chance of collisions between the molecules of the reactants decreases as time elapses. Hence, the rate of the reaction also decreases. The significance of this is that it shows that the rate of a reaction is proportional to the concentrations of the reactants. For the "generic" reaction $A + B \longrightarrow C$, this can be represented by the rate equation: Rate = $k[A]^a[B]^b$, where [A] and [B] are the concentrations (in moles/liter) of A and B respectively, a and b are the exponents to which those concentrations are raised, and k is the proportionality constant, usually called the rate constant.

The exponents a and b and the rate constant k can only be determined by doing experiments -- they have nothing to do with the stoichiometry of the reaction. Doing this kind of experiment can be challenging, however, since the rate of the reaction decreases as time elapses. One way to get around this problem is to measure the rate of the reaction as soon as the reactants are combined, before the rate has a chance to begin decreasing. This is known as the method of initial rates, since the rate is measured during the initial few minutes of the reaction. Measuring the initial rate for several different concentrations of reactants allows for the determination of the exponents in the rate equation. Once the exponents are known, the rate constant can be found by simple arithmetic, using the rate data.

THE METHOD OF INITIAL RATES:

Problem: The rate equation for the reaction: C_5H_{10} + HCl ➡ $C_5H_{11}Cl$ can be written as: Rate = $k[C_5H_{10}]^a[HCl]^b$. Find the values of the exponents a and b and the rate constant k. The data from several experiments are given below.[1]

Expt.	$[C_5H_{10}]$, M	[HCl], M	Initial Rate, M/sec
#1	0.00920	0.0322	1.76×10^{-5}
#2	0.0203	0.0329	3.96×10^{-5}
#3	0.0618	0.0329	1.20×10^{-4}
#4	0.0877	0.0153	3.92×10^{-5}
#5	0.0872	0.0271	1.22×10^{-4}
#6	0.0918	0.00352	2.07×10^{-6}

Solution: Compare experiments #1 and #2: [HCl] is (roughly) constant, but $[C_5H_{10}]$ goes up by a factor of 2.2. The rate also goes up by a factor of 2.2. In experiments #2 and #3, [HCl] is again constant, but $[C_5H_{10}]$ and the initial rate both go up by a factor of 3.0. Since whatever happens to $[C_5H_{10}]$ happens to the rate, a = 1. (Another way: $(2.2)^a = 2.2$. $(3.0)^a = 3.0$. Therefore, a = 1.)

Now compare experiments #4 and #5: $[C_5H_{10}]$ is (roughly) constant, but [HCl] goes up by a factor of 1.77. The initial rate goes up by a factor of 3.11 -- roughly the square of 1.77. In experiments #4 and #6, $[C_5H_{10}]$ is again (roughly) constant, but [HCl] goes down by a factor of 4.35. The initial rate goes down by a factor of 18.9 -- the square of 4.35. Whatever happens to [HCl] happens squared to the rate, so b = 2. (Another way: $(1.77)^b = 3.11$. $(4.35)^b = 18.9$. Thus, b = 2.)

Now that we know a and b, we can find k using data from experiment #1:

$$k = \text{Rate}/[C_5H_{10}][HCl]^2 = (1.76 \times 10^{-5}\ M\ sec^{-1}) \div [(.00920\ M)(.0322\ M)^2]$$

$k = 1.85\ M^{-2}\ sec^{-1}$. (Read as "1.85 per molar squared per second".)

1) Data taken from Y. Pocker, K. D. Stevens, and J. J. Champoux, Journal of the American Chemical Society, vol. 91 (1969), p. 4199.

ORDER OF A REACTION:

The <u>order</u> of a reaction is simply the sum of the exponents in the rate equation for that reaction. Thus, the reaction C_5H_{10} + HCl → $C_5H_{11}Cl$ is a <u>third-order reaction</u>, since its rate equation is: Rate = $k[C_5H_{10}]^1[HCl]^2$. (This reaction is said to be first-order with respect to C_5H_{10}, second-order with respect to HCl, and third-order overall, since 1 + 2 = 3.)

<u>First-order reactions</u> are those whose rate equations have the general form: Rate = $k[R]^1$, where [R] refers to the molar concentration of the reactant. It is possible to use integral calculus to convert this rate equation into another equation which more clearly shows how the concentration of the reactant in a first-order reaction changes as time elapses. We will not go through the calculus here (although readers who understand integral calculus are invited to try to obtain the equation immediately below from the rate equation above!); rather, we will simply show the equation that results. The new equation is: $kt = \ln([R]_i/[R]_f)$, where k is the <u>first-order rate constant</u> for the reaction, t is the amount of <u>time</u> that has elapsed, $[R]_i$ is the <u>initial</u> concentration of the reactant, and $[R]_f$ is the <u>final</u> concentration of the reactant -- that is, the concentration of the reactant after the time represented by t has elapsed. (The symbol "ln" stands for the <u>natural logarithm</u> of the quantity inside the parentheses. Refer to the appendix if you need to refresh your memory (or learn for the first time!) about logarithms.)

<u>Second-order reactions</u> are those whose rate equations have the general form: Rate = $k[R]^2$. Again, integral calculus may be used to convert this equation into the form which more clearly shows how the concentration of the reactant varies as a function of time. Again, we will not do the calculus here, but simply show the new equation for a second-order reaction: $kt = (1/[R]_f) - (1/[R]_i)$, where k is the <u>second-order rate constant</u>, and the other symbols are the same as before.

HALF-LIFE OF A REACTION:

You've probably heard the term "half-life" before in connection with radioactive decay. For example, the radioisotope radon-222 undergoes radioactive decay with a half-life of about four days. This means that half of any given sample of radon-222 will undergo radioactive decay over a four-day period of time. In general, the half-life of a reaction is the amount of time necessary for the concentration of a reactant to be reduced to one-half of its initial value. (In equation form, $[R]_f = [R]_i/2$ if $t = t_{1/2}$ ("$t_{1/2}$" is the symbol for the half-life).)

Given the above information, it is possible to derive expressions which relate the half-life of a reaction to the reaction's rate constant, using the time-concentration relationships. For example, for a first-order reaction:

$kt = \ln([R]_i/[R]_f)$ but if $t = t_{1/2}$, $[R]_f = [R]_i/2$. Therefore:

$kt_{1/2} = \ln([R]_i/([R]_i/2)) = \ln(2[R]_i/[R]_i) = \ln 2 = 0.693.$

Thus, $t_{1/2} = \underline{0.693/k}$ for any first-order reaction.

Notice that the half-life for a first-order reaction is a constant -- it depends only on the rate constant for the reaction, not on the reactant's concentration. This is one reason it is useful in discussing radioactive decay processes, all of which obey first-order kinetics. For a second-order reaction:

$kt = (1/[R]_f) - (1/[R]_i)$ but if $t = t_{1/2}$, $[R]_f = [R]_i/2$. Thus:

$kt_{1/2} = (1/([R]_i/2) - (1/[R]_i) = (2/[R]_i) - (1/[R]_i) = 1/[R]_i.$

Therefore, $t_{1/2} = \underline{1/k[R]_i}$ for any second-order reaction.

Problem: Calculate the half-lives of (a) a first-order reaction with a rate constant of 0.010 sec^{-1}, (b) a second-order reaction with a rate constant of 0.0214 M^{-1} sec^{-1} and an initial reactant concentration of 0.100 M.

Solution: (a) $t_{1/2} = 0.693/k = 0.693/(0.010\ sec^{-1})$ = 69 seconds.
(b) $t_{1/2} = 1/k[R]_i = 1/[(0.0214\ M^{-1}\ sec^{-1})(0.100\ M)]$ = 467 seconds.

ENERGY OF ACTIVATION:

The burning of wood is simply the reaction of the wood with oxygen in the air. Trees are surrounded by air, but they don't burst into flame unless the temperature around them is increased. For wood to burn, oxygen molecules must collide with the molecules in the wood. These collisions must be sufficiently forceful (that is, they must have enough energy) and must have the molecules in the correct orientation in order for covalent bonds to be broken and new covalent bonds to be formed. The minimum energy that a collision between molecules must possess in order to produce a reaction is called the energy of activation of that reaction.

At low temperatures, molecules move slowly. Collisions between slowly-moving molecules don't have enough energy to react, and the molecules just bounce off each other. At high temperatures, molecules move more rapidly -- that is, they have more kinetic energy. When two rapidly-moving molecules collide, their kinetic energy is transformed into potential energy at the instant the collision occurs. If the potential energy of the molecules at that instant is greater than the energy of activation, the reaction proceeds to give the products by the breaking and forming of covalent bonds. For such a collision, the instant of collision (where the molecules are "stuck together" and the bond-making and bond-breaking processes are in progress) is called the activated complex or the transition state.

In general, increasing the temperature of a reaction by about 10 ^{o}C increases the rate of the reaction by a factor of about three. A more precise correlation between the rate of a reaction, its temperature, and its energy of activation is given by the Arrhenius equation: $\ln k = \ln A - E_a/RT$. Here, k is the rate constant, E_a is the energy of activation, T is the Kelvin temperature, R is the gas constant expressed in energy units (1.987 cal/K mole or 8.314 J/K mole), and A is a "frequency factor" which represents how frequently molecules collide.

MECHANISM OF A REACTION:

Some reactions (such as the "S_N2" reaction: CH_3Br + OH^- ⟶ CH_3OH + Br^-) occur in a single step -- the reactants collide, and the products are formed. Other reactions (such as the "S_N1" reaction: C_4H_9Br + OH^- ⟶ C_4H_9OH + Br^-) occur in a series of steps, as shown in the sequence below:

<u>Step #1</u>: $C_4H_9Br \longrightarrow C_4H_9^+ + Br^-$

<u>Step #2</u>: $C_4H_9^+ + OH^- \longrightarrow C_4H_9OH$

A sequence of individual "steps" (or <u>elementary processes</u>, as the "steps" are often called) which describes the way in which a reaction takes place on the molecular level is called the <u>mechanism</u> of the reaction.

It is not possible to prove that a reaction proceeds by a particular mechanism, since we can't "see" molecules as they react. However, support for individual mechanisms can be found in the results of kinetics experiments. For example, consider the reactions above. The "S_N2" reaction has the following rate equation: Rate = $k[CH_3Br][OH^-]$. On the other hand, the "S_N1" reaction has this rate equation: Rate = $k[C_4H_9Br]$. Notice that the rate of the "S_N2" reaction depends on the concentration of OH^-, but the rate of the "S_N1" reaction does not. This is due to the fact that the "S_N1" reaction involves a <u>two-step</u> mechanism, in which the first step (above) occurs much more <u>slowly</u> than the second step (above). The <u>slowest</u> step in a reaction mechanism is called the <u>rate-determining step</u>, since the rate at which this step occurs determines the rate at which the entire reaction occurs. (As an analogy, consider several workers on an assembly line. No matter how fast the other workers work, the amount of work done by the <u>slowest</u> worker determines the rate at which the finished product is produced.) Since OH^- is not involved in the rate-determining step, the rate of the "S_N1" reaction does not depend on $[OH^-]$. Hence, Rate = $k[C_4H_9Br]$, as confirmed by kinetics experiments.

FREE-RADICAL "CHAIN" REACTIONS:

Ozone (O_3) is a form of oxygen which is formed in the upper atmosphere by the breakdown of O_2 into individual atoms of oxygen. These individual oxygen atoms then combine with molecules of O_2 to form molecules of O_3, as shown below:

Step #1: $O_2 + \text{energy} \rightarrow 2\,O$

Step #2: $O + O_2 \rightarrow O_3$

Since atmospheric ozone absorbs harmful ultraviolet radiation from the sun, thereby protecting us from skin cancer, it is important that the earth's ozone layer remain intact. Thus, there has been a growing concern since 1974 about the destruction of atmospheric ozone by a class of molecules called chlorofluorocarbons (or CFC's). CFC's are used in refrigerators, air conditioners, and aerosol spray cans. The problem with CFC's is that they tend to form free radicals upon exposure to ultraviolet light. Free radicals are atoms or groups of atoms which have at least one unpaired electron. Free radicals tend to be very reactive, since the unpaired electron can lower its energy by finding another electron and forming an electron pair. The sequence below shows how a typical CFC destroys atmospheric ozone:[1]

Step #1: $CFCl_3 + \text{U.V. light} \rightarrow CFCl_2 + :\ddot{\underset{..}{Cl}}\cdot$ (initiation)

Step #2: $:\ddot{\underset{..}{Cl}}\cdot + O_3 \rightarrow O_2 + :\ddot{\underset{..}{Cl}}-\ddot{\underset{..}{O}}\cdot$ (propagation)

Step #3: $:\ddot{\underset{..}{Cl}}-\ddot{\underset{..}{O}}\cdot + O \rightarrow O_2 + :\ddot{\underset{..}{Cl}}\cdot$ (propagation)

This process is called a free-radical chain reaction, since chlorine free radicals are formed in Step #1 (the initiation step) and then react with ozone in Step #2 and Step #3 (the chain propagation steps). Notice that at the end of Step #3, the chlorine radical generated in Step #1 has been re-generated, and is thus available to destroy another ozone molecule. Since Steps #2 and #3 are repeated many times (hence the "chain" reaction), one CFC molecule can destroy 1000 molecules of O_3.

---

1) TIME Magazine, October 19, 1987 issue.

HOW CATALYSTS WORK:

Catalysts are substances which affect the rate of chemical reactions without being consumed themselves during the course of the reaction. Usually, the effect of a catalyst is to increase the rate of a reaction. This happens because the presence of the catalyst provides an alternative mechanism by which a reaction can proceed. The alternative mechanism usually has a lower energy of activation than the uncatalyzed mechanism. Therefore, the reaction will proceed at a faster rate, since a greater percentage of the collisions between molecules will have a sufficient amount of energy to produce a reaction and lead to formation of product.

There are two main types of catalysts: homogeneous and heterogeneous. Homogeneous catalysts are contained in the same phase as the rest of the reaction mixture during the course of the reaction. An example is sulfuric acid, which catalyzes the reaction between acetic acid and amyl alcohol to form amyl acetate (pear oil). Sulfuric acid, acetic acid, and amyl alcohol form a homogeneous mixture, and the purpose of the sulfuric acid is to react with the acetic acid and make it more susceptible to attack by molecules of amyl alcohol. If desired, the sulfuric acid can be recovered, apparently unreacted, at the end of the reaction. (Sulfuric acid does react, but is re-generated during the reaction.)

Heterogeneous catalysts occupy a different phase from the reactants during the course of the reaction. Typical heterogeneous catalysts include the transition metals nickel, palladium, and platinum. These catalysts are used in the reaction of hydrogen with organic materials in a process called hydrogenation. You're probably most familiar with this process as a way to change vegetable oils into vegetable shortening -- the only difference is a few hydrogen atoms in each molecule! Molecules of H_2 and vegetable oils become attached to the metal surface, which assists in the bond-making and bond-breaking processes in the reaction.

DYNAMIC EQUILIBRIUM AND LeCHATELIER'S PRINCIPLE:

During the course of the "generic" reaction R $\rightarrow$ P, the concentration of the reactants gradually decreases, and the concentration of the products slowly increases. Since the concentration of the reactants is decreasing, the rate of the reaction will also decrease. However, since the concentration of the products is increasing, the rate of the reverse reaction (P $\rightarrow$ R) will tend to increase if the energy of activation of the reverse reaction is not too large to prevent this. As the rate of the forward reaction (R $\rightarrow$ P) decreases and the rate of the reverse reaction increases, a point will be reached at which both reactions are proceeding at the same rate. This situation is called dynamic equilibrium -- the system is at equilibrium (since the concentrations of reactants and products do not change), but individual molecules continue to undergo reactions and are therefore dynamic. ("Dynamic" here means "moving" or "changing".)

Dynamic equilibrium is a relatively low energy state for a system. Indeed, the free energy (ΔG) of a system at equilibrium is zero. Most systems try to minimize their energy content whenever possible. Therefore, dynamic equilibrium is a desirable state for a system. If something happens to a system at equilibrium to disturb the system and remove it from its state of equilibrium, then the system will do whatever it can to return to its equilibrium state. This principle was first stated in 1888 by the French physical chemist Henri Louis LeChatelier, and it has come to be known as LeChatelier's Principle.

For the equilibrium reaction rR $\rightleftharpoons$ pP (the double arrows represent equilibrium), the mass action expression (sometimes called the reaction quotient, Q) is written as: $[P]^p/[R]^r$. (Note that the coefficients in the balanced equation equal the exponents in the reaction quotient.) The system is at equilibrium when the reaction quotient equals the equilibrium constant (K_{eq}) for that reaction.

FREE ENERGY AND EQUILIBRIUM CONSTANTS:

Ammonia (NH_3) can be formed from nitrogen and hydrogen by the reaction: $N_{2\,(g)} + 3\ H_{2\,(g)} \rightleftharpoons 2\ NH_{3\,(g)}$. This reaction is called the Haber Process.

Problem: What is the mass action expression for this equilibrium?

Solution: If you remember that the rule is "products-over-reactants", and that the coefficients in the balanced equation equal the exponents in the mass action expression, this should be easy: $[NH_3]^2/[N_2][H_2]^3$.

Problem: A 1.00-liter flask contains 8.2 moles of NH_3, 0.100 moles of N_2, and 0.100 moles of H_2 when the system has come to equilibrium. Calculate the equilibrium constant for the above reaction.

Solution: Since the volume of the flask is 1.00 liter, the amounts given above are numerically the same as the concentrations of each substance. Thus:

$[NH_3] = 8.2\ M \qquad [N_2] = 0.100\ M \qquad [H_2] = 0.100\ M$

$K_{eq} = [NH_3]^2/[N_2][H_2]^3 = (8.2\ M)^2/(0.100\ M)(0.100\ M)^3$

$K_{eq} = 6.7 \times 10^5$.[1] (Units are usually omitted for equilibrium constants.)

Problem: Calculate ΔG^o_r for the above reaction, given that ΔG^o_r and K_{eq} are related by the following equation: $\Delta G^o_r = -RT \ln K_{eq}$.

Solution: This should be easy if you recall that the "o" in "ΔG^o_r" refers to a temperature of 25.0 ^oC, which equals 298 K. Therefore:

$$\Delta G^o_r = -RT \ln K_{eq} = -(1.987\ cal/K\ mole)(298\ K) \ln(6.7 \times 10^5)$$

$$\Delta G^o_r = -(1.987\ cal/K\ mole)(298\ K)(13.42) = -7.95 \times 10^3\ cal/mole.$$ [1]

This reaction is spontaneous at 25 ^oC. The spontaneity of the reaction is also exemplified by the large value of the equilibrium constant. Large values of K_{eq} imply spontaneous reactions (favoring formation of products); small values of K_{eq} imply non-spontaneous reactions (favoring formation of the reactants).

---

1) CRC Handbook of Chemistry and Physics, 56th Edition, 1975-1976.

USING LeCHATELIER'S PRINCIPLE:

The Haber Process reaction, $N_{2\ (g)} + 3\ H_{2\ (g)} \rightleftharpoons 2\ NH_{3\ (g)}$, is exothermic (its ΔH_r^o value is -22.08 kcal/mol[1]). LeChatelier's Principle says that if a system at equilibrium is disturbed, either the forward reaction or the reverse reaction will occur predominantly until the equilibrium is restored.

Problem: If $N_{2\ (g)}$ is added to the equilibrium, what will happen?

Solution: Adding N_2 increases $[N_2]$, which increases the rate of the forward reaction. Thus, the forward reaction will occur -- NH_3 will be formed, and N_2 and H_2 will be consumed -- until equilibrium is restored. (This is sometimes summarized by saying that the equilibrium "shifts to the right", towards products.)

Problem: If $NH_{3\ (g)}$ is added to the equilibrium, what will happen?

Solution: Adding NH_3 increases $[NH_3]$, which increases the rate of the reverse reaction. Thus, the reverse reaction will occur -- NH_3 will be consumed, and N_2 and H_2 will be formed -- until equilibrium is restored. (This is sometimes summarized by saying that the equilibrium "shifts to the left", towards reactants.)

Problem: If heat is added to the equilibrium, what will happen?

Solution: Think of it this way -- for an exothermic reaction, heat is a product. Therefore, the situation is similar to adding NH_3 to the system -- the equilibrium will shift to the left. (If the reaction had been endothermic, adding heat would shift the equilibrium to the right. This is van't Hoff's Law.)

Problem: If the container is compressed, what will happen?

Solution: Compressing the container increases all three concentrations. To restore the equilibrium, the system will shift in the direction which allows it to reduce the total number of gas particles. In this case, the shift is to the right, since four moles of gas are on the left, but only two moles are on the right.

---

1) CRC Handbook of Chemistry and Physics, 56th Edition, 1975-76.

SOME "REAL WORLD" EFFECTS OF DYNAMIC EQUILIBRIA:

Dynamic equilibria affect us in many ways. Most notable are the many biological and biochemical processes that depend on dynamic equilibria. As an example, consider the following equilibrium:

$$\underset{\text{(hydroxyapatite)}}{Ca_{10}(PO_4)_6(OH)_2} + 2\ F^- \rightleftharpoons \underset{\text{(fluoroapatite)}}{Ca_{10}(PO_4)_6F_2} + 2\ OH^-$$

Hydroxyapatite and fluoroapatite are the mineral forms of calcium found in human bones and teeth. "Fluoroapatite is less easily dissolved by mouth acids than hydroxyapatite and therefore more resistant to decay."[1] Increasing the amount of fluoride ions present in the bloodstream shifts the above equilibrium to the right, favoring the formation of the acid-resistant fluoroapatite. This is one reason why many dentists advise drinking fluoridated water and using toothpaste which contains fluoride ions. The fluoridation of water remains controversial, as some people are adversely affected by an excess of fluoride in their bloodstreams.[1]

Another example involves the tendency of chickens to lay eggs with shells that are too thin to survive handling and processing. This usually occurs in hot weather, when chickens are more likely to pant to rid themselves of excess body heat. "Panting can appreciably lower the level of carbon dioxide in a chicken's blood."[2] This shifts the following equilibrium to the left:

$$H_2O_{(l)} + CO_{2\,(g)} \rightleftharpoons 2\ H^+_{(aq)} + CO_3^{2-}{}_{(aq)}$$

"The shift reduces the amount of carbonate (CO_3^{2-}) ion available to the chicken for making eggshells, which are 95% calcium carbonate ($CaCO_3$) in the form of calcite. Poultry farmers could solve the problem by having their flocks drink Perrier water or just plain carbonated water."[2] Experiments support this claim.[3]

---

1) Chemical & Engineering News, August 1, 1988, page 29.
2) Chemical & Engineering News, March 20, 1989, page 94.
3) Odun, T. W., et al., Poultry Science, vol. 64 (1985), page 594, as cited in Chemical & Engineering News, June 12, 1989, page 48.

SOLVING EQUILIBRIUM PROBLEMS:

Problem: Consider the equilibrium: $PCl_{5\,(g)} \rightleftharpoons PCl_{3\,(g)} + Cl_{2\,(g)}$. The equilibrium constant for this reaction is 7.30×10^{-2}. If 1.00 mole of PCl_5 is injected into an empty 1.00-liter container, what will be the concentrations of PCl_5, PCl_3, and Cl_2 when the system has come to equilibrium?[1]

Solution: First, write the equilibrium expression for this reaction:

$K_{eq} = [PCl_3][Cl_2]/[PCl_5] = 7.30 \times 10^{-2}$.

Now, during the reaction, some PCl_5 will decompose, forming PCl_3 and Cl_2. What we don't know is how much PCl_5 will react in this way. Hence, call the amount of PCl_5 that reacts "X". From the balanced equation above, if "X" moles of PCl_5 react, "X" moles of PCl_3 and "X" moles of Cl_2 will be formed. These results are summarized in the following table:

	Initial Value	Change During Reaction	Final Value
$[PCl_5]$	1.00 M	-X	(1.00 - X) M
$[PCl_3]$	0.00 M	+X	X M
$[Cl_2]$	0.00 M	+X	X M

Inserting the "Final Value" entries into the equilibrium expression above gives:

$$[PCl_3][Cl_2]/[PCl_5] = 7.30 \times 10^{-2} = (X)(X)/(1.00 - X) = X^2/(1.00 - X).$$

This equation can be rewritten in the form of a quadratic equation, as shown below:

$$X^2 = (7.30 \times 10^{-2})(1.00 - X) = 7.30 \times 10^{-2} - (7.30 \times 10^{-2})(X)$$

$$X^2 + (7.30 \times 10^{-2})(X) - 7.30 \times 10^{-2} = 0 = aX^2 + bX + c$$

where $a = 1$, $b = 7.30 \times 10^{-2}$, and $c = -7.30 \times 10^{-2}$. Solving the quadratic equation:

$$X = [-b \pm (b^2 - 4ac)^{1/2}]/2a$$

$$X = [-(7.30 \times 10^{-2}) + ((7.30 \times 10^{-2})^2 - (4)(1)(-7.30 \times 10^{-2}))^{1/2}]/(2)(1) = \underline{0.236}.$$

Thus: $[PCl_5] = 1.00 - 0.236 = \underline{0.76\ M}$. $[PCl_3] = \underline{0.236\ M}$. $[Cl_2] = \underline{0.236\ M}$.

---

1) From a similar problem on page 313 in "Chemical Principles and Properties" by M. J. Sienko and R. A. Plane. 1974 by McGraw-Hill, Inc.

AN ALTERNATIVE METHOD FOR SOLVING SOME EQUILIBRIUM PROBLEMS:

Problem: The equilibrium constant is 1.8×10^{-6} for the equilibrium $2\ NO_{2\,(g)} \rightleftharpoons 2\ NO_{(g)} + O_{2\,(g)}$. If 1.00 mole of NO_2 is injected into an empty 1.00-liter container, what will be the concentrations of NO_2, NO, and O_2 when the system has come to equilibrium?[1]

Solution: First, write the equilibrium expression for this reaction:

$$K_{eq} = [NO]^2[O_2]/[NO_2]^2 = 1.8 \times 10^{-6}.$$

During the reaction, if "X" moles of O_2 are formed, "2X" moles of NO will be formed, and "2X" moles of NO_2 will be consumed. See the table below:

	Initial Value	Change During Reaction	Final Value
$[NO_2]$	1.00 M	-2X	(1.00 - 2X) M
[NO]	0.00 M	+2X	2X M
$[O_2]$	0.00 M	+X	X M

Inserting the "Final Value" entries into the equilibrium expression above gives:

$$[NO]^2[O_2]/[NO_2]^2 = 1.8 \times 10^{-6} = (2X)^2(X)/(1.00 - 2X)^2 = 4X^3/(1.00-2X)^2.$$

Solving this cubic equation would be quite challenging! However, there's a "short cut" that we can take here because the equilibrium constant is a very small number. This implies that the forward reaction does not proceed very far -- that is, that "X" will also be a very small number. If "X" is very small, then the "2X" in the denominator above can be neglected with respect to the "1.00" from which it is being subtracted. Hence, the denominator is approximately equal to 1.00. This makes solving the above equation much easier, as shown below:

$$4X^3/(1.00)^2 = 1.8 \times 10^{-6} = 4X^3. \quad X^3 = (1.8 \times 10^{-6})/4 = 4.5 \times 10^{-7}.$$

$$X = (4.5 \times 10^{-7})^{1/3} = \underline{7.7 \times 10^{-3}} = [O_2] \text{ at equilibrium.}$$

Thus: $[NO] = (2)(7.7 \times 10^{-3}) = \underline{1.5 \times 10^{-2}\ M}$. $[NO_2] = 1.00 - 0.015 = \underline{0.98\ M}$.

---

1) From a similar problem on page 473 in "General Chemistry" by R. H. Petrucci. Fourth Edition, 1985 by Macmillan Publishing Company.

HETEROGENEOUS EQUILIBRIA:

A heterogeneous equilibrium is one in which the species in equilibrium are present in different physical states or phases. An example is the equilibrium $C_{(s)} + O_{2\,(g)} \rightleftharpoons CO_{2\,(g)}$, in which solid carbon and two gases are all in equilibrium. You might think that the equilibrium expression for this equilibrium would be written as: $K_{eq} = [CO_2]/[C][O_2]$. However, unlike the concentrations of gases (which can be compressed) or solutions (which have variable compositions), the "concentrations" of pure solids and pure liquids are constants, since they only depend on the density and molecular weight (both constants) of the solid or liquid. Hence, the "concentration" term for a pure solid or pure liquid is incorporated as part of the equilibrium constant, and the equilibrium expressions of heterogeneous equilibria do not include these terms. Hence, the "[C]" term should not appear in the equilibrium expression above, and the correct way to write this equilibrium expression is: $K_{eq} = [CO_2]/[O_2]$. In a similar fashion, the equilibrium expression for the heterogeneous equilibrium $2\,H_{2\,(g)} + O_{2\,(g)} \rightleftharpoons 2\,H_2O_{(l)}$ should be written as: $K_{eq} = 1/[H_2]^2[O_2]$, since $H_2O_{(l)}$ is a pure liquid.

A saturated solution is prepared by adding solute to a solvent until no more solute will dissolve, and then adding extra solute, which sinks to the bottom of the container. In this system, the undissolved solute is in equilibrium with the dissolved solute. This is therefore a heterogeneous equilibrium. For example, the equilibrium $BaSO_{4\,(s)} \rightleftharpoons Ba^{2+}_{(aq)} + SO_4^{2-}{}_{(aq)}$ is present in a saturated solution of barium sulfate. Since $[BaSO_4]$ is a constant, it is omitted from the equilibrium expression, and the correct form of the equilibrium expression is: $K_{eq} = [Ba^{2+}][SO_4^{2-}]$. Since this is simply the product of the concentrations of the ions in solution, it is usually called the solubility product, and the equilibrium constant is called the solubility product constant, or K_{sp} for short.

USING SOLUBILITY PRODUCTS:

Problem: The concentration of Ba^{2+} ions in a saturated solution of $BaSO_4$ is 1.04×10^{-5} M. What is the solubility product constant (K_{sp}) for $BaSO_4$?

Solution: The equilibrium is: $BaSO_{4\,(s)} \rightleftharpoons Ba^{2+}{}_{(aq)} + SO_4{}^{2-}{}_{(aq)}$. Thus, equal numbers of barium ions and sulfate ions must be present in solution. Therefore, $[SO_4{}^{2-}] = [Ba^{2+}]$, and the value of K_{sp} is easily calculated, as shown:

$$K_{sp} = [Ba^{2+}][SO_4{}^{2-}] = (1.04 \times 10^{-5})(1.04 \times 10^{-5}) = \underline{1.08 \times 10^{-10}}.$$ [1]

Problem: The K_{sp} value for silver chloride (AgCl) is 1.56×10^{-10}.[1] Calculate $[Ag^+]$ and $[Cl^-]$ in a saturated aqueous solution of AgCl.

Solution: The equilibrium is: $AgCl_{(s)} \rightleftharpoons Ag^+{}_{(aq)} + Cl^-{}_{(aq)}$. Call the amount of AgCl which dissolves "X". If "X" moles of AgCl dissolve, "X" moles of Ag^+ ions and "X" moles of Cl^- ions must be formed. See the table below:

	Initial Value	Change During Reaction	Final Value
$[Ag^+]$	0.00 M	+X	X M
$[Cl^-]$	0.00 M	+X	X M

Inserting the "Final Value" entries into the equilibrium expression gives:

$$K_{sp} = 1.56 \times 10^{-10} = [Ag^+][Cl^-] = (X)(X) = X^2.$$

$$X = (1.56 \times 10^{-10})^{1/2} = \underline{1.25 \times 10^{-5}} = [Ag^+] = [Cl^-].$$

Problem: The K_{sp} value for lead iodide (PbI_2) is 1.39×10^{-8}.[1] Calculate $[Pb^{2+}]$ and $[I^-]$ in a saturated aqueous solution of lead iodide.

Solution:

(The equilibrium is: $PbI_{2(s)} \rightleftharpoons Pb^{2+}{}_{(aq)} + 2\,I^-{}_{(aq)}$.)

	Initial	Change	Final
$[Pb^{2+}]$	0.00 M	+X	X M
$[I^-]$	0.00 M	+2X	2X M

$$K_{sp} = 1.39 \times 10^{-8} = [Pb^{2+}][I^-]^2 = (X)(2X)^2 = 4X^3.$$

$$X = (1.39 \times 10^{-8}/4)^{1/3} = \underline{1.51 \times 10^{-3}} = [Pb^{2+}]. \quad [I^-] = \underline{3.03 \times 10^{-3}\ M}.$$

1) CRC Handbook of Chemistry and Physics, 56th Edition, 1975-1976.

THE COMMON ION EFFECT:

The solubility of an ionic solute in water is <u>decreased</u> if the water already contains one of the ions which is present in the solute. This observation is called the <u>common ion effect</u>, and is illustrated by the following situation:

<u>Problem</u>: The K_{sp} value for lead iodide (PbI_2) is 1.39×10^{-8}.[1] Calculate $[Pb^{2+}]$ in a saturated aqueous solution of lead iodide if enough potassium iodide (KI) is added to make $[I^-] = 0.100$ M.

<u>Solution</u>: The equilibrium is: $PbI_{2\,(s)} \rightleftharpoons Pb^{2+}_{(aq)} + 2\,I^-_{(aq)}$. Call the amount of PbI_2 which dissolves "X". If "X" moles of PbI_2 dissolve, "X" moles of Pb^{2+} ions and "2X" moles of I^- ions will be formed. However, the <u>total</u> concentration of I^- ions in the solution is already known: $[I^-] = 0.100$ M. Thus:

	Initial Value	Change During Reaction	Final Value
$[Pb^{2+}]$	0.00 M	+X	X M
$[I^-]$	0.00 M	+2X	0.100 M

Inserting the "Final Value" entries into the equilibrium expression gives:

$$K_{sp} = 1.39 \times 10^{-8} = [Pb^{2+}][I^-]^2 = (X)(0.100)^2 = (X)(0.0100).$$

$$X = (1.39 \times 10^{-8})/(0.0100) = \underline{1.39 \times 10^{-6}} = [Pb^{2+}].$$

Notice that this concentration is <u>lower</u> than the concentration of Pb^{2+} in a saturated aqueous solution of PbI_2 <u>without</u> added iodide ion ($[Pb^{2+}] = 1.51 \times 10^{-3}$). This should not be surprising, since LeChatelier's Principle predicts that adding I^- ions to the above equilibrium should shift it to the <u>left</u>, <u>lowering</u> $[Pb^{2+}]$.

<u>Problem</u>: Calculate $[I^-]$ in a saturated aqueous solution of lead iodide if enough lead nitrate ($Pb(NO_3)_2$) is added to make $[Pb^{2+}] = 0.100$ M.

<u>Solution</u>: $K_{sp} = 1.39 \times 10^{-8} = [Pb^{2+}][I^-]^2 = (0.100)(X)^2$.

$$X = [(1.39 \times 10^{-8})/(0.100)]^{1/2} = (1.39 \times 10^{-7})^{1/2} = \underline{3.73 \times 10^{-4}} = [I^-].$$

---

1) CRC Handbook of Chemistry and Physics, 56th Edition, 1975-1976.

THE SELF-IONIZATION OF WATER:

Consider the equilibrium: $\mathrm{H{:}\overset{\large H}{\underset{..}{\overset{..}{O}}}{:}} + \mathrm{H{:}\overset{\large H}{\underset{..}{\overset{..}{O}}}{:}} \rightleftharpoons \mathrm{H{:}\overset{\large H}{\underset{..}{\overset{..}{O}}}{:}H} + \mathrm{H{:}\underset{..}{\overset{..}{O}}{:}}$, which can also be written as: $2\,H_2O_{(l)} \rightleftharpoons H_3O^+_{(aq)} + OH^-_{(aq)}$. Notice that the water molecule on the right transfers a proton to the water molecule on the left. (A hydrogen ion, H^+, is commonly called a proton, since that's what it is -- a hydrogen atom which has lost its only electron is simply a proton!) This occurs to a small extent in any system which contains water. The above equilibrium is a heterogeneous equilibrium, so the equilibrium expression does not include a $[H_2O]$ term. The equilibrium expression for the above equilibrium is: $[H_3O^+][OH^-] = K_w$, where K_w (the ion product constant for water, hence the "w") has a value of 1.00×10^{-14} at a temperature of 25 ^oC. The above equilibrium is sometimes called the self-ionization of water, since water molecules form ions when they interact.

The above equilibrium also illustrates several ways of looking at acid-base chemistry. Acids and bases have been defined in several ways, all of which can be useful. The Arrhenius definition describes acids as compounds which form H_3O^+ ions (called hydronium ions) when placed in water, and bases as compounds which form OH^- ions (called hydroxide ions) when placed in water. (Water itself is amphoteric -- that is, able to react as either an acid or a base -- since both H_3O^+ and OH^- ions are formed in the above reaction.) The Brønsted definition describes acids as proton donors, and bases as proton acceptors. In the above reaction, the water molecule on the right "donates" a proton to the water molecule on the left, which "accepts" the proton. (Again, water behaves as either an acid or a base.) The Lewis definition describes acids as electron-pair acceptors, and bases as electron-pair donors. In the above reaction, notice that the water molecule on the left "donates" one of the lone pairs of electrons on the oxygen atom to the proton which "accepts" it, thereby forming a coordinate covalent bond.

ACIDIC, BASIC, AND NEUTRAL SOLUTIONS:

Problem: Calculate $[H_3O^+]$ and $[OH^-]$ in pure water, given the equilibrium: $2\ H_2O_{(l)} \rightleftharpoons H_3O^+_{(aq)} + OH^-_{(aq)}$. $K_w = 1.00 \times 10^{-14}$.

Solution: Since the reaction of two moles of water molecules produces one mole of hydronium ions and one mole of hydroxide ions, $[H_3O^+] = [OH^-]$. Thus:

$$K_w = [H_3O^+][OH^-] = 1.00 \times 10^{-14} = (X)(X) = X^2.$$

$$X = (1.00 \times 10^{-14})^{1/2} = \underline{1.00 \times 10^{-7}} = [H_3O^+] = [OH^-].$$

Pure water is neutral, since $[H_3O^+] = [OH^-]$. In an acidic solution, $[H_3O^+]$ is greater than $[OH^-]$, and in a basic solution, $[OH^-]$ is greater than $[H_3O^+]$.

To avoid the extensive use of scientific notation when discussing the acidity or basicity of aqueous solutions, the terms "pH" and "pOH" are often used. The "p" refers to the negative logarithm of the quantity in question. Therefore, $pH = -\log[H_3O^+]$, and $pOH = -\log[OH^-]$. ("pH" $= -\log[H^+] = -\log[H_3O^+]$.)

Problem: Calculate the pH and the pOH of neutral water.

Solution: In neutral water, $[H_3O^+] = [OH^-] = 1.00 \times 10^{-7}$ M.

Therefore, $pH = pOH = -\log(1.00 \times 10^{-7}) = \underline{7.000}$.

Notice that pH + pOH = 7.000 + 7.000 = 14.000. The sum of the pH and the pOH of any aqueous solution will always be equal to 14.000 at 25 °C. (This should not be surprising, since 14.000 is simply "pK_w" -- the negative logarithm of K_w.)

In an acidic solution, the pH is less than 7.000. (This may be a bit confusing, since $[H_3O^+]$ is greater than 1.00×10^{-7} M in an acidic solution, but the negative sign in the definition of pH causes the trend to be reversed.) Since pH + pOH = 14.000, it follows that pOH is greater than 7.000 in an acidic solution. In a basic solution, exactly the opposite is true -- the pH is greater than 7.000, and the pOH is less than 7.000. Remember, though -- regardless of whether the solution is acidic, basic, or neutral, the sum of the pH and the pOH is 14.000.

USING pH AND pOH:

One way in which pH and pOH are typically used is to describe the acidity or basicity of <u>dilute</u> solutions of <u>strong</u> acids and <u>strong</u> bases. A <u>strong</u> acid or base is, by definition, <u>totally</u> dissociated into ions when placed in water. Therefore, for <u>strong</u> acids, [acid] = $[H_3O^+]$, and for <u>strong</u> bases, [base] = $[OH^-]$.

<u>Problem</u>: What is the pH of a 0.0100 M HCl solution?

<u>Solution</u>: HCl is a strong acid, so [HCl] = 1.00×10^{-2} M = $[H_3O^+]$.

Therefore, pH = $-\log[H_3O^+] = -\log(1.00 \times 10^{-2}) =$ <u>2.000</u>.

<u>Problem</u>: What is the pH of a 2.0×10^{-4} M NaOH solution?

<u>Solution</u>: NaOH dissolves completely in water to give Na^+ and OH^- ions. Therefore, [NaOH] = 2.0×10^{-4} M = $[OH^-]$. Now, using pH + pOH = 14.000, we get:

pOH = $-\log[OH^-] = -\log(2.0 \times 10^{-4}) = 3.70$.

pH = 14.000 - pH = 14.000 - 3.70 = <u>10.30</u>.

<u>Problem</u>: "Normally, rain has a pH of 5.6, but over much of the Northeast it's an acidic 4.3."[1] How many times more acidic is Northeastern rain than normal rain? (Rain which has a pH value of 5.0 or less is called "acid rain".)

<u>Solution</u>: One way to approach this problem is to convert the pH values above into $[H_3O^+]$ values and then divide the two $[H_3O^+]$ values. Hence:

Normal rain: $[H_3O^+] = 10^{-pH} = 10^{-5.6} =$ <u>2.5×10^{-6} M</u>.

Northeastern rain: $[H_3O^+] = 10^{-pH} = 10^{-4.3} =$ <u>5.0×10^{-5} M</u>.

$(5.0 \times 10^{-5}\ M)/(2.5 \times 10^{-6}\ M) =$ <u>20</u>. Therefore, Northeastern rain is <u>twenty times more acidic than normal rain</u>. This points out an important feature of using pH and pOH: due to the use of the logarithmic function, a change of <u>1.0 unit</u> of pH or pOH corresponds to a <u>tenfold</u> change in the acidity or basicity. (Another way to do the problem: subtract the pH values <u>first</u>. $10^{-(4.3 - 5.6)} = 10^{1.3} =$ <u>20</u>.)

---

1) U. S. News & World Report, July 25, 1988, page 44.

CALCULATING pH FOR SOLUTIONS OF WEAK ACIDS:

One way in which pH is typically used is to describe the acidity of solutions of weak acids. A weak acid, by definition, is not completely dissociated into ions when placed into water. Thus, for weak acids, [acid] $\neq$ [H_3O^+]. Problems such as this must therefore be solved using techniques for dealing with equilibria. The equilibrium constant for the reaction in which a weak acid dissociates into ions has a particular symbol -- K_a, the acid dissociation constant.

Problem: Calculate the pH of a 0.100 M solution of acetic acid (the acid which is a major component of vinegar.). $K_a = 1.76 \times 10^{-5}$ for acetic acid.[1]

Solution: The equilibrium for this reaction is given below:

$$HC_2H_3O_{2\,(aq)} + H_2O_{(l)} \rightleftharpoons H_3O^+_{(aq)} + C_2H_3O_2^-{}_{(aq)}$$

This is a heterogeneous equilibrium, so no [H_2O] term appears in the equilibrium expression. The equilibrium expression is: $K_a = [H_3O^+][C_2H_3O_2^-]/[HC_2H_3O_2]$. Since we don't know how much of the acetic acid ($HC_2H_3O_2$) dissociates, call that amount "X". If "X" moles of acetic acid dissociates, "X" moles of H_3O^+ and "X" moles of $C_2H_3O_2^-$ will be formed. This information is summarized in the following table:

	Initial Value	Change During Reaction	Final Value
[$HC_2H_3O_2$]	0.100 M	-X	(0.100 - X) M
[$C_2H_3O_2^-$]	0.000 M	+X	X M
[H_3O^+]	0.000 M	+X	X M

Weak acids have small K_a values. This tells us that "X" will be a small number, negligible with respect to 0.100. Hence, using the information above, we get:

$$K_a = [H_3O^+][C_2H_3O_2^-]/[HC_2H_3O_2] = 1.76 \times 10^{-5} = (X)(X)/0.100 = X^2/0.100.$$

$$X = [(1.76 \times 10^{-5})(0.100)]^{1/2} = (1.76 \times 10^{-6})^{1/2} = 1.33 \times 10^{-3} = [H_3O^+].$$

$$pH = -\log[H_3O^+] = -\log(1.33 \times 10^{-3}) = \underline{2.877}.$$

1) CRC Handbook of Chemistry and Physics, 56th Edition, 1975-1976.

CALCULATING pH FOR SOLUTIONS OF WEAK BASES:

One way in which pH is typically used is to describe the basicity of solutions of <u>weak</u> bases. A <u>weak</u> base, by definition, does <u>not</u> completely react to form OH^- ions when placed in water. Thus, for <u>weak</u> bases, [base] $\neq$ $[OH^-]$. In this case, the problem must be solved using techniques for dealing with <u>equilibria</u>. The equilibrium constant for the reaction in which a weak base reacts to form OH^- ions has a particular symbol -- <u>K_b</u>, the <u>base dissociation constant</u>.

<u>Problem</u>: Calculate the pH of a 0.100 M solution of <u>ammonia</u> (yes, the same substance used as a household cleaner!). $K_b = 1.79 \times 10^{-5}$ for ammonia.[1]

<u>Solution</u>: The equilibrium for this reaction is given below:

$$NH_{3\,(aq)} + H_2O_{(l)} \rightleftharpoons NH_4^+{}_{(aq)} + OH^-{}_{(aq)}$$

This is a <u>heterogeneous</u> equilibrium, so no $[H_2O]$ term appears in the equilibrium expression. The equilibrium expression is: $K_b = [NH_4^+][OH^-]/[NH_3]$. Since we don't know how much of the ammonia (NH_3) reacts, call that amount "X". If "X" moles of ammonia reacts, "X" moles of NH_4^+ and "X" moles of OH^- will be formed.

Summarizing:	Initial Value	Change During Reaction	Final Value
$[NH_3]$	0.100 M	-X	(0.100 - X) M
$[NH_4^+]$	0.000 M	+X	X M
$[OH^-]$	0.000 M	+X	X M

The <u>small</u> value of K_b (typical for a <u>weak</u> base) tells us that "X" is a small number and is negligible compared to 0.100. Hence, using the information above, we get:

$$K_b = [NH_4^+][OH^-]/[NH_3] = 1.79 \times 10^{-5} = (X)(X)/0.100 = X^2/0.100.$$

$$X = [(1.79 \times 10^{-5})(0.100)]^{1/2} = (1.79 \times 10^{-6})^{1/2} = 1.34 \times 10^{-3} = [OH^-].$$

$$pOH = -\log[OH^-] = -\log(1.34 \times 10^{-3}) = 2.874.$$

$$pH = 14.000 - pOH = 14.000 - 2.874 = \underline{11.126}.$$

---

1) CRC Handbook of Chemistry and Physics, 56th Edition, 1975-1976.

CONJUGATE ACIDS AND CONJUGATE BASES:

Consider the equilibrium for the reaction of a "generic" weak acid, HA, with water: $HA_{(aq)} + H_2O_{(l)} \rightleftharpoons H_3O^+_{(aq)} + A^-_{(aq)}$. If the reaction was to proceed in the reverse direction, the effect of the A^- ion would be to accept a proton (H^+) from the H_3O^+ ion -- that is, the A^- ion would be effectively functioning as a base. For this reason, A^- is referred to as the conjugate base of the acid HA. For a specific example, consider the water molecule in the above reaction -- it accepts a proton from HA, and forms H_3O^+ in the process. Therefore, H_2O can be considered the conjugate base of the acid H_3O^+.

Similarly, consider the equilibrium for the reaction of the A^- ion with water: $A^-_{(aq)} + H_2O_{(l)} \rightleftharpoons HA_{(aq)} + OH^-_{(aq)}$. Again, reversing this reaction shows that HA would donate a proton to the OH^- ion, thus functioning as an acid. Therefore, HA is referred to as the conjugate acid of the base A^-. Again, a specific example would be the water molecule above -- it donates a proton to the A^- ion, forming OH^-. Therefore, H_2O is the conjugate acid of the base OH^-. (Notice that the difference between an acid or base and its conjugate is just one proton!)

The equilibrium expressions for the above equilibria are given below:

$K_a = [H_3O^+][A^-]/[HA]$ $\qquad$ $K_b = [HA][OH^-]/[A^-]$

When these two expressions are multiplied together, something interesting happens:

$$K_aK_b = ([H_3O^+][A^-]/[HA]) \times ([HA][OH^-]/[A^-]) = [H_3O^+][OH^-] = K_w.$$

Thus, for a conjugate acid-base pair (that is, an acid and its conjugate base or a base and its conjugate acid), $K_a \times K_b = K_w = 1.00 \times 10^{-14}$. From this equation, it can be seen that if K_a is large, K_b must be small, and if K_b is large, K_a must be small -- that is, the stronger the acid, the weaker its conjugate base, and the stronger the base, the weaker its conjugate acid. Acids and bases tend to react in such a way as to form their weaker conjugates whenever possible.

CALCULATING pH FOR SOLUTIONS OF SALTS OF WEAK ACIDS:

<u>Problem</u>: Acetic acid ($HC_2H_3O_2$) has a K_a value of 1.76×10^{-5}.[1] Calculate the pH of a 1.00 M solution of sodium acetate ($NaC_2H_3O_2$) in water.

<u>Solution</u>: When sodium acetate dissolves in water, sodium ions (Na^+) and acetate ions ($C_2H_3O_2^-$) are formed. The acetate ions react with water as shown by the equilibrium: $C_2H_3O_2^-{}_{(aq)} + H_2O_{(l)} \rightleftarrows HC_2H_3O_2{}_{(aq)} + OH^-{}_{(aq)}$. Since OH^- ions are formed in this reaction, the solution will be basic, and its pH will be greater than 7.000. The value of K_b for the acetate ion can be calculated from the equation $K_aK_b = K_w$, since $C_2H_3O_2^-$ is the conjugate base of $HC_2H_3O_2$. Hence:

$$K_b = K_w/K_a = (1.00 \times 10^{-14})/(1.76 \times 10^{-5}) = 5.68 \times 10^{-10}.$$

The equilibrium expression for the above equilibrium is given below:

$$K_b = [HC_2H_3O_2][OH^-]/[C_2H_3O_2^-] = 5.68 \times 10^{-10}$$

To find the concentration of OH^-, the usual methods for dealing with equilibria are used. Call the amount of acetate ion that reacts "X". If "X" moles of acetate ion react, "X" moles of acetic acid and "X" moles of hydroxide ion will be formed.

Summarizing:

	Initial Value	Change During Reaction	Final Value
$[C_2H_3O_2^-]$	1.00 M	-X	(1.00 - X) M
$[HC_2H_3O_2]$	0.00 M	+X	X M
$[OH^-]$	0.00 M	+X	X M

Inserting the "Final Value" entries into the above equilibrium expression gives the following (the "X" in the denominator is negligible, due to the small value of K_b):

$$K_b = [HC_2H_3O_2][OH^-]/[C_2H_3O_2^-] = 5.68 \times 10^{-10} = (X)(X)/(1.00) = X^2.$$

$$X = (5.68 \times 10^{-10})^{1/2} = 2.38 \times 10^{-5} = [OH^-].$$

$$pOH = -\log[OH^-] = -\log(2.38 \times 10^{-5}) = 4.623.$$

$$pH = 14.000 - pOH = 14.000 - 4.623 = \underline{9.377}. \text{ (Greater than 7.000!)}$$

---

1) CRC Handbook of Chemistry and Physics, 56th Edition, 1975-1976.

BUFFER SOLUTIONS:

Living things cannot tolerate drastic changes in pH in their habitats. For example, in a recent study supported by the Canadian government, "researchers dumped sulfuric acid into a small Ontario lake, systematically raising its acid level 30 times to a pH of 5. When they were done, one third of the resident species had been knocked out -- including the shrimp and crayfish eaten by trout. The emaciated fish stopped reproducing. Relates team leader David Schindler: 'We eventually would have lost every species of fish.'"[1]

Fortunately, oceans and lakes are natural buffer solutions. A <u>buffer solution</u> is one which contains both a weak Brønsted acid <u>and</u> its conjugate base. It is therefore able to resist changes in pH by reacting with either acids or bases which are added to the solution. If an acid is added to a buffer solution, the base already present reacts with it; if a base is added to a buffer solution, the acid already present reacts with it.

The pH of a buffer solution depends upon the concentrations of the acid and conjugate base used to prepare it. In order to see this, consider the generic weak acid equilibrium: $HA_{(aq)} + H_2O_{(l)} \rightleftharpoons H_3O^+_{(aq)} + A^-_{(aq)}$. The equilibrium expression for this equilibrium is: $K_a = [H_3O^+][A^-]/[HA]$. Now, if we take the negative logarithm of each side of this equation, we get the following:

$$-\log K_a = -\log([H_3O^+][A^-]/[HA]) = -\log[H_3O^+] - \log([A^-]/[HA])$$

However, since "$-\log[H_3O^+]$" is simply <u>pH</u>, and "$-\log K_a$" is called <u>pK_a</u>, we get:

$$pK_a = pH - \log([A^-]/[HA]) \quad \text{or, as it is usually written:}$$

$$\underline{pH = pK_a + \log([A^-]/[HA])}.$$

The underlined equation is known as the <u>Henderson-Hasselbalch Equation</u>, and gives us an easy way to find the pH of a buffer solution if pK_a, [HA], and $[A^-]$ are known.

---

1) U. S. News & World Report, July 25, 1988, page 44.

HOW BUFFERS WORK:

Problem: Calculate the pH of a buffer solution made by dissolving 1.00 mole of sodium acetate and 0.100 mole of acetic acid in enough water to make 1.00 liter of solution. Acetic acid has a K_a value of 1.76×10^{-5}.[1]

Solution: Since the volume of the solution is 1.00 liter, $[HC_2H_3O_2]$ = 0.100 M and $[C_2H_3O_2^-]$ = 1.00 M. Now, using the Henderson-Hasselbalch Equation:

$$pH = pK_a + \log([A^-]/[HA]) = -\log K_a + \log([C_2H_3O_2^-]/[HC_2H_3O_2]).$$

$$pH = -\log(1.76 \times 10^{-5}) + \log(1.00\ M/0.100\ M) = 4.754 + 1.000 = \underline{5.754}.$$

Problem: Calculate the pH of the buffer solution formed by adding 0.100 mole of HCl to the above buffer solution.

Solution: HCl is a strong acid, so the effect of adding HCl to the buffer solution is to increase the concentration of H_3O^+. This shifts the equilibrium $HC_2H_3O_{2\,(aq)} + H_2O_{(l)} \rightleftharpoons H_3O^+_{(aq)} + C_2H_3O_2^-{}_{(aq)}$ to the left, increasing $[HC_2H_3O_2]$ and decreasing $[C_2H_3O_2^-]$. If we assume that all of the HCl reacted with acetate ion to form acetic acid, $[HC_2H_3O_2]$ should increase by 0.100 M, and $[C_2H_3O_2^-]$ should decrease by 0.100 M. Therefore, at equilibrium:

$$[HC_2H_3O_2] = 0.100\ M + 0.100\ M = \underline{0.200\ M}.$$

$$[C_2H_3O_2^-] = 1.00\ M - 0.100\ M = \underline{0.90\ M}.$$

Now, simply apply the Henderson-Hasselbalch Equation, as in the previous problem:

$$pH = pK_a + \log([A^-]/[HA]) = -\log K_a + \log([C_2H_3O_2^-]/[HC_2H_3O_2])$$

$$pH = -\log(1.76 \times 10^{-5}) + \log(0.90\ M/0.200\ M) = 4.754 + 0.65 = \underline{5.41}.$$

Notice that the addition of the HCl only changes the pH of the buffer solution from 5.754 to 5.41. This corresponds to about a twofold increase in acidity. If the HCl had been added to neutral (unbuffered) water, the pH would have changed from 7.000 to 1.000 -- a millionfold increase in acidity! Buffers resist changes in pH.

---

1) CRC Handbook of Chemistry and Physics, 56th Edition, 1975-1976.

POLYPROTIC ACIDS AND THEIR SALTS:

A polyprotic acid is an acid whose molecules are each capable of donating more than one proton. Examples of polyprotic acids include sulfuric acid (H_2SO_4), carbonic acid (H_2CO_3), and phosphoric acid (H_3PO_4). Since more than one proton can be donated by each of these acids, more than one equilibrium expression is needed to describe the reactions that take place. For example, the equilibria below describe the reactions of carbonic acid (a diprotic acid) with water:

$$H_2CO_{3\,(aq)} + H_2O_{(l)} \rightleftharpoons H_3O^+_{(aq)} + HCO_3^-{}_{(aq)} \qquad K_{a1} = 4.30 \times 10^{-7}$$

$$HCO_3^-{}_{(aq)} + H_2O_{(l)} \rightleftharpoons H_3O^+_{(aq)} + CO_3^{2-}{}_{(aq)} \qquad K_{a2} = 5.61 \times 10^{-11}$$

The following equilibria describe the reactions of H_3PO_4 (a triprotic acid):

$$H_3PO_{4\,(aq)} + H_2O_{(l)} \rightleftharpoons H_3O^+_{(aq)} + H_2PO_4^-{}_{(aq)} \qquad K_{a1} = 7.52 \times 10^{-3}$$

$$H_2PO_4^-{}_{(aq)} + H_2O_{(l)} \rightleftharpoons H_3O^+_{(aq)} + HPO_4^{2-}{}_{(aq)} \qquad K_{a2} = 6.23 \times 10^{-8}$$

$$HPO_4^{2-}{}_{(aq)} + H_2O_{(l)} \rightleftharpoons H_3O^+_{(aq)} + PO_4^{3-}{}_{(aq)} \qquad K_{a3} = 2.2 \times 10^{-13}$$

The K_a values[1] shown are typical for polyprotic acids. From the K_a values, it can be seen that the first proton is donated more readily than the other protons are. (K_{a1} is larger than K_{a2}, which in turn is larger than K_{a3}.) This is due to the fact that the first proton donated leaves a neutral molecule, whereas the second proton donated leaves a negative ion. Protons, being positively charged, are less likely to leave negative ions than neutral molecules. The large differences between the successive K_a values imply that all of the molecules donate their first protons before any of them donate their second or third protons.

The acid salts of polyprotic acids are excellent buffers, since they can either donate protons or accept protons. (Phosphate and carbonate buffers keep our blood at a relatively constant pH of about 7.4.) Buffers like this can be made simply by titrating a polyprotic acid with NaOH until the desired pH is obtained.

1) CRC Handbook of Chemistry and Physics, 56th Edition, 1975-1976.

HYDROLYSIS OF SALTS:

When table salt (sodium chloride, NaCl) is dissolved in water, the pH of the resulting solution is <u>neutral</u> -- 7.000. However, other salts form solutions which are not neutral. For example, sodium acetate ($NaC_2H_3O_2$) gives a <u>basic</u> solution when dissolved in water, and ammonium chloride (NH_4Cl) produces an <u>acidic</u> solution. This comes about due to the reaction of water molecules with one of the ions of the salt, producing OH^- ions or H_3O^+ ions. This process is called <u>hydrolysis</u>, which comes from "hydro" (meaning <u>water</u>) and "lysis" (meaning <u>cleavage</u>).

The tendency of some ions to cause hydrolysis to occur is predictable. Ions which are the <u>conjugate bases of weak acids</u> tend to produce <u>basic</u> solutions. Examples include the phosphate ion (PO_4^{3-}), the carbonate ion (CO_3^{2-}), and the acetate ion ($C_2H_3O_2^-$). Ions which are the <u>conjugate acids of weak bases</u> tend to produce <u>acidic</u> solutions. An example is the ammonium ion (NH_4^+). It is noteworthy that two of the above ions (phosphate and carbonate) tend to form ionic compounds which are insoluble in water, but soluble in acids. The basicity of these ions is the property that causes them both to produce hydrolysis and to dissolve in acids. Ions whose ionic compounds are generally water-soluble (such as Na^+, Cl^-, K^+, and NO_3^-) do <u>not</u> cause hydrolysis to occur. They are sometimes called "spectator ions", since they do not react with water. It is noteworthy that the "spectator ions" Cl^- and NO_3^- are the conjugate bases of <u>strong</u> acids. Since <u>strong</u> acids have rather <u>weak</u> conjugate bases, these ions tend not to react with water very strongly.

<u>Problem</u>: Predict the acidity or basicity of: (a) NH_4NO_3, (b) K_2CO_3.

<u>Solution</u>: (a) is acidic, (b) is basic. One way to approach this problem (other than by looking at the information above) is to notice that NH_4NO_3 is formed by the reaction of HNO_3 (a <u>strong</u> acid) and NH_3 (a <u>weak</u> base). Therefore, the solution will be <u>acidic</u>. KOH (<u>strong</u> base) + H_2CO_3 (<u>weak</u> acid) $\longrightarrow$ K_2CO_3.

OXIDATION-REDUCTION REACTIONS:

The rusting of iron and the burning of coal are two common examples of oxidation-reduction reactions (or, for short, "redox" reactions). It is not too surprising that the reaction of something with oxygen is an "oxidation" reaction, but what is "reduction"? To find out, let's look at the above reactions:

$$2\ :\ddot{\underset{\cdot\cdot}{Fe}}:\ +\ 3\ :\dot{\underset{\cdot}{O}}:\ \longrightarrow\ 2\ :\dot{\underset{\cdot\cdot}{Fe}}{}^{3+}\ +\ 3\ :\ddot{\underset{\cdot\cdot}{O}}:^{2-}\ \longrightarrow\ Fe_2O_3.$$

$$:\dot{C}\cdot\ +\ 2\ :\dot{\underset{\cdot}{O}}:\ \longrightarrow\ :\underset{\cdot\cdot}{O}::C::\underset{\cdot\cdot}{O}:\ =\ CO_2.$$

In the reaction between iron atoms and oxygen atoms, electrons are transferred from the iron atoms to the oxygen atoms. This transfer of electrons from one atom to another is oxidation and reduction in the purest sense. The loss of electrons by an atom is called oxidation; when an atom gains electrons, the process is called reduction. Thus, in the first reaction above, iron atoms are oxidized, and oxygen atoms are reduced. The two processes of oxidation and reduction always take place simultaneously -- if something is oxidized, something else must be reduced. Hence, for an oxidation to take place, something must do the oxidizing. The atom which accepts electrons from another atom, thereby oxidizing it, is called the oxidizing agent. Not surprisingly, oxygen is the oxidizing agent in the reactions above. Similarly, for a reduction to occur, something must do the reducing. The atom that donates electrons to another atom, thereby reducing it, is called the reducing agent. Iron is the reducing agent in the first reaction above.

In the second reaction above, electrons are shared by the carbon and oxygen atoms. However, since oxygen atoms are more electronegative than carbon atoms, we can think of the shared electrons as more the "property" of the oxygen atoms than the carbon atom. Hence, electrons are "transferred", and this is an oxidation-reduction reaction. Notice that the reducing agent (here, carbon) is, itself, oxidized, while the oxidizing agent (here, oxygen) is, itself, reduced.

OXIDATION NUMBERS:

An easy way to keep track of the transferring of electrons (or electron density, for covalent molecules) that occurs during oxidation-reduction reactions is to assign oxidation numbers to each atom present. An atom's oxidation number (or oxidation state) is simply the electrical charge that the atom would have if the compound in which that atom is located were completely ionic.

For ionic compounds, assigning oxidation numbers is easy -- just use the charges on the ions themselves! For example, the ionic compound Fe_2O_3 is made up of Fe^{3+} ions and O^{2-} ions. Therefore, in Fe_2O_3, the oxidation number of iron is +3, and oxygen is in the -2 oxidation state. For covalent compounds, assigning oxidation numbers is a little bit more challenging, since no ions are present. However, the process can be simplified by simply considering all electrons present in covalent bonds to be the "property" of the more electronegative atom in the bond. In other words, assign oxidation numbers to atoms in covalent molecules as if the compound were completely ionic. For example, consider the covalent molecule CO_2, whose Lewis Structure is $:\ddot{O}::C::\ddot{O}:$ (Lone pairs of electrons are the "property" of the atom on which they are located.). If all four electrons in each double bond are assigned to the oxygen atoms, then each oxygen atom will "own" eight electrons. This is the electron configuration for the O^{2-} ion, so oxygen's oxidation number is -2. The carbon atom will have lost all four of its valence electrons, so it will be in the +4 oxidation state. A good way to check to see if you have assigned the oxidation numbers correctly is to see if the sum of the oxidation numbers is equal to the total charge on the formula as written. For Fe_2O_3, a neutral formula, the sum should equal zero, and it does: $(+3) + (+3) + (-2) + (-2) + (-2) = 0$. For CO_2, another neutral molecule, the sum is: $(+4) + (-2) + (-2) = 0$. For the polyatomic ion MnO_4^-, Mn is in the +7 oxidation state: $(+7) + (-2) + (-2) + (-2) + (-2) = -1$.

GALVANIC CELLS:

Different kinds of batteries are used as energy sources in many of the electrically-based instruments we use routinely. Batteries provide energy for our cars, our flashlights, our wristwatches, and our calculators. The energy that a battery supplies comes from chemical reactions that take place inside the battery. Batteries are sometimes called galvanic cells (or voltaic cells). They differ from electrolytic cells in that galvanic cells use a chemical reaction (usually a redox reaction) to supply electrical energy, whereas exactly the opposite is true in an electrolytic cell -- electrical energy is used to make a chemical reaction occur.

In a typical galvanic cell, the electrodes and solutions for each of the two "half-reactions" are kept separate from each other. (This is in contrast to the situation in a typical electrolysis cell, in which the electrodes and the solutions are often in direct contact with each other.) The only connection between the two "half-reactions" is usually a wire which connects the electrodes. Therefore, for the transfer of electrons from one "half-reaction" to the other to occur, the electrons must flow through the wire connecting the electrodes. The flow of electrons through a wire is commonly called an electric current. Thus, an oxidation-reduction reaction can be used to generate a current of electricity.

The force which "pushes" electrons through a wire is called the electromotive force (or e.m.f., for short). This force is measured in units called volts. (1 volt = 1 Joule/1 coulomb.) Therefore, if a voltmeter is connected to each of the two electrodes, the voltage (sometimes called the potential difference) between the two "half-reactions" can be measured. This voltage is sometimes called the cell potential (symbol E_{cell}) of a galvanic cell. The standard cell potential (symbol E^0_{cell}) of a galvanic cell is the cell potential of the galvanic cell when measured at 25 ^{0}C, 1.00 atmosphere, and concentrations of all ions equal to 1.00 M.

STANDARD REDUCTION POTENTIALS:

It is not possible to measure the voltage produced by a single "half-reaction", since oxidation and reduction always happen simultaneously. When faced with a situation like this, scientists do the next best thing -- they declare one "half-reaction" to be the standard against which all other "half-reactions" are measured. The standard hydrogen electrode (SHE), for which the "half-reaction" is $2\ H^+_{(aq)} + 2\ e^- \rightleftharpoons H_{2\ (g)}$, consists of a platinum electrode surrounded by hydrogen gas and immersed in a solution which contains H^+ ions -- that is, an acid solution. The potential of this "half-reaction" has been arbitrarily declared to be 0.0000 volts.[1] The potential of any other "half-reaction" can thus be measured by connecting its electrode to the SHE through a voltmeter. The reading on the voltmeter is the voltage (or potential difference) between the two electrodes, and is called the standard reduction potential (symbol E^o) for the "half-reaction" in question if the measurement is made at 25 ^oC, 1.00 atm, and [ion] = 1.00 M.

Notice that the standard reduction potential is written as a reduction, with the ions on the left gaining electrons. Notice also that it is written as an equilibrium, since in any given galvanic cell, any given "half-reaction" may be either the oxidation or the reduction. To find the standard cell potential of a particular galvanic cell, simply find the difference between the standard reduction potentials of the two "half-reactions" in the galvanic cell. (That's why the voltage is sometimes called the "potential difference"!) In equation form, this is: $E^o_{cell} = E^o_{reduction} - E^o_{oxidation}$. This equation can also be used to tell which "half-reaction" is the oxidation and which "half-reaction" is the reduction, since all functioning galvanic cells have a positive value of E^o_{cell}. If the subtraction above produces a negative E^o_{cell} value, the cathode and anode should be reversed.

1) CRC Handbook of Chemistry and Physics, 56th Edition, 1975-1976.

USING STANDARD REDUCTION POTENTIALS:

<u>Problem</u>: Given the following standard reduction potentials[1]:

$Cu^{2+} + 2e^- \rightleftharpoons Cu \qquad E^o = +0.3402$ V

$Zn^{2+} + 2e^- \rightleftharpoons Zn \qquad E^o = -0.7628$ V

calculate E^o_{cell} for the following cell: Zn / Zn^{2+} (1.00 M) // Cu^{2+} (1.00 M) / Cu.

<u>Solution</u>: The notation above is sometimes used to describe galvanic cells. The advantage of this notation is that it makes it somewhat easier to see the half-reactions that occur in the cell. The half-reactions can be understood by simply reading from left to right -- Zn is converted into Zn^{2+}, and Cu^{2+} becomes Cu. Thus, the overall cell reaction is: $Zn_{(s)} + Cu^{2+}_{(aq)} \rightarrow Zn^{2+}_{(aq)} + Cu_{(s)}$. Since zinc is being <u>oxidized</u>, zinc is the <u>anode</u> in this cell; since copper is being <u>reduced</u>, copper is the <u>cathode</u> in this cell. To find E^o_{cell}, just subtract the two standard reduction potentials above, following the equation below:

$$E^o_{cell} = E^o_{reduction} - E^o_{oxidation} = E^o_{cathode} - E^o_{anode} = E^o_{Cu} - E^o_{Zn}.$$

$$E^o_{cell} = (+0.3402 \text{ V}) - (-0.7628 \text{ V}) = \underline{+1.103 \text{ volts}}.$$

<u>Problem</u>: Given the following standard reduction potentials[1]:

$Fe^{2+} + 2e^- \rightleftharpoons Fe \qquad E^o = -0.409$ V

$Cr^{3+} + 3e^- \rightleftharpoons Cr \qquad E^o = -0.74$ V

calculate E^o_{cell} for the following cell: Fe / Fe^{2+} (1.00 M) // Cr^{3+} (1.00 M) / Cr.

<u>Solution</u>: It's tempting to try to make some kind of "correction" for the fact that different numbers of electrons are being transferred in the two half-reactions, but don't do this -- just find the potential difference, as above:

$$E^o_{cell} = E^o_{redn.} - E^o_{oxdn.} = E^o_{Cr} - E^o_{Fe} = (-0.74 \text{ V}) - (-0.409 \text{ V}) = \underline{-0.33 \text{ V}}.$$

The <u>negative</u> value of E^o_{cell} tells us that the cell notation is <u>incorrect</u>. Actually, chromium is the <u>anode</u>, iron is the <u>cathode</u>, and the real value of E^o_{cell} is <u>+0.33 V</u>.

---

1) CRC Handbook of Chemistry and Physics, 56th Edition, 1975-1976.

APPLICATIONS OF GALVANIC CELLS:

People use many different types of batteries every day. One of the most common types of batteries is the car battery, also known as the lead storage battery. The electrodes in the lead storage battery consist of metallic lead and lead(IV) oxide, each of which is converted into lead sulfate as the battery runs. The cell notation for this battery is: $Pb_{(s)}$ / $PbSO_{4\,(s)}$ // $PbO_{2\,(s)}$ / $PbSO_{4\,(s)}$. The source of the sulfate ions is a dilute solution of sulfuric acid that serves as the electrolyte in this battery. The fluid nature of the electrolyte allows the ions in solution to move freely, which enables this battery to be recharged. In a similar fashion, the nickel-cadmium batteries used in calculators may be recharged, since the electrolyte present is a solution of hydroxide ions. In these batteries, the electrodes are metallic cadmium and and nickel(IV) oxide. The cell notation for this battery is: $Cd_{(s)}$ / $Cd(OH)_{2\,(s)}$ // $NiO_{2\,(s)}$ / $Ni(OH)_{2\,(s)}$.

By contrast, common flashlight batteries cannot be recharged. These batteries are sometimes called zinc-carbon dry cells, and their "dryness" is what prevents them from being recharged. The electrodes in the zinc-carbon dry cell are the zinc casing of the battery and the rod of carbon that extends through the center of the battery. The space between the electrodes is filled with a thick paste that contains manganese(IV) oxide, among other things. The cell notation for this battery is: Zn / Zn^{2+} // MnO_2 / Mn^{3+}. The Zn^{2+} and Mn^{3+} ions are held in place by the thick paste, and are not free to move. This hinders any efforts to recharge a flashlight battery, since in order to make the redox reaction run in the reverse direction, an external source of electricity is used to reverse the polarity of the electrodes. The ions must be free to migrate to the opposite electrode from the electrode at which they are formed for the reverse reaction to proceed. This is not the case in the zinc-carbon "dry" cell.

ELECTROCHEMISTRY:

Sodium is a highly reactive metal. Chlorine is a poisonous, greenish-yellow gas. When sodium and chlorine react, the relatively stable compound sodium chloride (table salt) is formed. The reaction is: $2\ Na + Cl_2 \rightarrow 2\ NaCl$. This is an oxidation-reduction reaction -- sodium is oxidized, chlorine is reduced. The reaction proceeds in the direction shown -- the reverse reaction does not occur spontaneously. (In fact, metallic sodium does not occur in nature. Sodium is so reactive that all naturally-occurring sodium exists in the form of sodium salts.) However, it is possible to make the reverse of the above reaction occur by taking electrons away from the Cl^- ions and bringing them into contact with the Na^+ ions. This is done by using a current of electricity as the source of electrons. In a typical Downs cell (used for the industrial production of sodium metal and chlorine gas), sodium chloride is heated until it melts. Two electrodes are then immersed into the molten sodium chloride. The electrodes are connected to a source of electrical current, and a current of electricity is allowed to flow through the electrodes and the molten sodium chloride. (For the current to flow, the ions must be free to move around, which explains why the sodium chloride is melted.) The Na^+ ions migrate toward the negative electrode (called the <u>cathode</u>, because <u>cat</u>ions are attracted to it) and are reduced by the electrons produced by the electric current. The Cl^- ions migrate toward the positive electrode (called the <u>anode</u>, since <u>an</u>ions are attracted to it) and donate electrons to it, thereby becoming oxidized. The reactions which occur during the <u>electrolysis</u> of sodium chloride are shown below:

Cathode: $Na^+ + e^- \rightarrow Na$ (reduction)

Anode: $2\ Cl^- \rightarrow 2\ e^- + Cl_2$ (oxidation)

This process illustrates one aspect of <u>electrochemistry</u>, which is the study of the relationships between electrical energy and chemical (usually <u>redox</u>) reactions.

QUANTITATIVE ELECTROLYSIS:

Electrolysis (from "electro-", meaning electricity, and "-lysis", which means cleavage or breaking apart) is the process by which a compound is converted into the elements from which it is made by using an electrical current to perform oxidation-reduction reactions. One of the most practical uses of electrolysis is in the production of metals from their ions. In order to find practical conditions for doing this, some of the quantitative aspects of electrolysis must be considered.

The unit of electrical current is the ampere. One ampere (or amp, for short) is the electrical current caused by having one coulomb of electrical charge pass through a wire in one second. In equation form, this can be written as either 1 amp = 1 coul/sec or 1 coulomb = (1 amp)(1 sec). Since the particles which carry electrical charges while a current flows are electrons, and since the electrical charge on an individual electron is known to be 1.60×10^{-19} coulombs, we can write the following expression for the charge on one mole (6.02×10^{23}) of electrons: (6.02×10^{23} electrons/mole)(1.60×10^{-19} coul/electron) = 96,500 coulombs/mole e^-. This number is sometimes called Faraday's constant (or one Faraday, symbol F) in honor of the English physicist Michael Faraday, who discovered electromagnetics.

Problem: What mass of copper metal is produced by passing a current of 1.00 amp for 5.00 minutes through a solution of Cu^{2+} ions?

Solution: Since 1 coulomb = (1 amp)(1 sec), the electrical charge is:

(1.00 amp)(5.00 min)(60 sec/min) = 300 (amp x sec) = 300 coulombs.

Now, since 96,500 coulombs = 1 mole of electrons, the amount of electrons is:

(300 coulombs)(1 mole of electrons/96,500 coulombs) = 3.11×10^{-3} moles.

The "half-reaction" which occurs is: $Cu^{2+}_{(aq)} + 2e^- \rightarrow Cu_{(s)}$. Hence:

(3.11×10^{-3} moles e^-)(1 mole Cu/2 moles e^-) = 1.55×10^{-3} moles Cu.

(1.55×10^{-3} moles Cu)(63.54 g/mole Cu) = 0.0988 grams of Cu.

APPLICATIONS OF ELECTROLYSIS:

The process of electrolysis is useful for producing many things that we use routinely, or that affect our lives in other ways. One of the most notable of these is the production or purification of various metals by electrolysis. Sodium metal is produced from molten sodium chloride by electrolysis in the Downs cell. (A "cell" is the term used to describe any electrochemical process, or the vessel in which the process occurs.) The Downs process also produces chlorine gas, which can be used in the production of pesticides. The sodium chloride used in the Downs process comes from an abundant source: evaporation of sea water! Sea water also contains magnesium ions, and magnesium metal can be produced by electrolysis of $MgCl_2$ in a process called the Dow process. Copper and aluminum metals can also be produced in this way, although the Hall process for producing aluminum is expensive. (It's cheaper to recycle aluminum cans than to mine and produce new aluminum!)

A variation on this basic theme is the process known as electroplating, in which thin metal coatings are applied to a metal surface. This is done simply by setting up an electrolysis cell, using the metal surface to be coated as the cathode and immersing it in a solution which contains ions of the metal to be used as the coating. "Gold-plated" jewelry, "silver" eating utensils, and the "chrome" finish on automobile grillwork ("chrome" is short for chromium) are all produced by this basic process.

Pure water does not conduct electricity, but sea water contains enough ions to allow it to conduct electricity quite well. Electrolysis of sea water produces three major products: chlorine gas, hydrogen gas (used in the manufacture of fertilizers), and sodium hydroxide (used in making soaps). The reactions are:

Anode: $2\ Cl^-_{(aq)} \longrightarrow Cl_{2\ (g)} + 2\ e^-$ (oxidation)

Cathode: $2\ H_2O_{(l)} + 2\ e^- \longrightarrow H_{2\ (g)} + 2\ OH^-_{(aq)}$ (reduction)

NUCLEAR CHEMISTRY:

In the latter part of the twentieth century, the word "nuclear" has come to be associated with catastrophic events, such as "nuclear war" or "nuclear winter". This is unfortunate, because the word "nuclear" is derived from the word "nucleus", and is intended simply to describe things associated with atomic nuclei. Therefore, "nuclear chemistry" is simply the chemistry of the atomic nucleus. Most of the chemical reactions that atoms undergo involve the transfer or sharing of one or more of the atoms' valence electrons, but this is not the case for nuclear reactions, which involve only the nuclei of atoms.

One unusual feature of most nuclear reactions is that atoms of one element are often changed into atoms of another element. The transformation of an atom of one element into an atom of another element is called transmutation. One way in which transmutation can occur is for an unstable atomic nucleus to stabilize itself by giving off radiation in the form of small particles of matter or photons of energy. An atomic nucleus which gives off radiation of this type is said to be radioactive. The radioactivity of an atomic nucleus is its relative tendency to give off this kind of radiation. ("Radioactivity" also means the radiation itself.)

Some of the particles of matter typically emitted by radioactive nuclei are listed below. The emission process is sometimes called "radioactive decay", since a large atomic nucleus "decays" in the process of forming smaller nuclei.

Name of Particle	Symbol(s) of Particle	Description
Alpha Particle	${}^{4}_{2}He$, ${}^{4}_{2}He^{2+}$, or ${}^{4}_{2}\alpha$	helium nucleus
Beta Particle	${}^{0}_{-1}e$, ${}^{0}_{-1}e^{-}$, or ${}^{0}_{-1}\beta$	electron
Gamma Particle	${}^{0}_{0}\gamma$	photon of energy
Positron	${}^{0}_{1}e$, ${}^{0}_{1}e^{+}$, or ${}^{0}_{1}\beta^{+}$	positive electron
Neutron	${}^{1}_{0}n$	neutron

BALANCING EQUATIONS OF NUCLEAR REACTIONS:

The total mass of the products of a nuclear reaction is equal to the total mass of the reactants. Similarly, the total charge of the products of a nuclear reaction is equal to the total charge of the reactants. Therefore, to write a balanced equation for a nuclear reaction, all that is necessary is to be sure that the sum of the mass numbers (superscripts) is the same for the reactants as it is for the products, and that the sum of the atomic numbers (subscripts) is the same for the reactants as it is for the products. Examples are given below.

Problem: A uranium-238 nucleus undergoes radioactive decay, becoming a nucleus of thorium-234. Write a balanced equation for this nuclear reaction.

Solution: The difference between the mass numbers of the reactant and product is 4 (238 - 234). Hence, it is likely that the emitted particle is an alpha particle, since an alpha particle has a mass number of 4. This is confirmed by the fact that the difference in atomic numbers between uranium (atomic number = 92) and thorium (atomic number = 90) is 2 (92 - 90). This agrees with the atomic number of an alpha particle (a helium nucleus), which is also 2. Therefore, the balanced equation for this nuclear reaction is: ${}^{238}_{92}U \rightarrow {}^{234}_{90}Th + {}^{4}_{2}\alpha$.

Problem: A thorium-234 nucleus undergoes radioactive decay, giving off a beta particle and a gamma particle. Write a balanced equation for this reaction.

Solution: Beta particles and gamma particles both have mass numbers of zero, so no change will occur in the atom's mass number -- it will remain at 234. Gamma particles have atomic numbers of zero, but the "atomic number" of a beta particle is -1. (Beta particles are electrons, each of which carries a -1 charge.) Therefore, the atomic number of the product nucleus must be the difference between the atomic numbers of thorium (90) and a beta particle (-1). 90 - (-1) = 91, the atomic number of protactinium. The equation is: ${}^{234}_{90}Th \rightarrow {}^{234}_{91}Pa + {}^{0}_{-1}\beta + {}^{0}_{0}\gamma$.

NUCLEAR STABILITY:

Atomic nuclei which are radioactive give off energy in order to become more stable. The stability of atomic nuclei depends upon several factors. Since atomic nuclei contain positively-charged protons which repel each other, neutrons stabilize the nucleus by separating the protons and "insulating" them from each other. The greater the number of protons present in an atomic nucleus, the greater the number of neutrons needed for the nucleus to be stable. The ratio of protons to neutrons in stable nuclei is about 1:1 for relatively small nuclei, but is about 2:3 for larger, more massive nuclei. In addition, nuclei that contain <u>even</u> numbers of protons or neutrons are generally more stable than nuclei that contain <u>odd</u> numbers of protons or neutrons. (Of the 264 known stable nuclei, 157 contain <u>even</u> numbers of protons <u>and</u> neutrons, but only 5 contain <u>odd</u> numbers of protons and neutrons.) Also, nuclei which contain certain "magic numbers" of protons or neutrons tend to be more stable than other nuclei. The "magic numbers" are 2, 8, 20, 50, 82, and 126. Nobody knows why these numbers are "magic" for atomic nuclei.

Atomic nuclei which contain too many protons <u>and</u> neutrons to be stable often emit <u>alpha</u> particles. Alpha particles contain two protons and two neutrons each, so emission of an alpha particle considerably reduces the mass of a nucleus. Atomic nuclei which contain too many neutrons to be stable usually emit <u>beta</u> particles. Beta particles are electrons, but there are no electrons in an atomic nucleus. However, when a nucleus emits a beta particle, one neutron in the nucleus is changed into a proton: ${}^{1}_{0}n \longrightarrow {}^{0}_{-1}\beta + {}^{1}_{1}H$. (The symbol for a proton is ${}^{1}_{1}H$.) Atomic nuclei which contain too many protons to be stable usually emit <u>positrons</u>. Positrons are particles the size of electrons, but having a <u>positive</u> charge. When a positron collides with an electron, <u>both</u> are destroyed. Emission of a positron changes a proton in the nucleus into a neutron: ${}^{1}_{1}H \longrightarrow {}^{0}_{1}e^{+} + {}^{1}_{0}n$.

BINDING ENERGY:

One way to measure the relative stability of an atomic nucleus is to calculate its <u>binding energy</u>. The mass of any atomic nucleus is <u>less</u> than the sum of the masses of the protons and neutrons which make it up. The difference between the two masses is called the binding energy, and it can be expressed in units of energy by using Einstein's well-known equation: $E = mc^2$, where m is the mass of a particle, E is the energy which corresponds to that mass (here, the binding energy), and c is the speed of light (3.00×10^8 m/sec). The greater the binding energy -- that is, the greater the amount of energy <u>lost</u> by the nucleus in the process of constructing it from protons and neutrons -- the more stable the nucleus.

<u>Problem</u>: Calculate the binding energy of a nucleus of carbon-14, given the following information[1]: Mass of ^{14}C Nucleus = 14.007682 amu. Mass of Proton = 1.008142 amu. Mass of Neutron = 1.008982 amu. 1 amu = 1.66×10^{-24} grams.

<u>Solution</u>: A $^{14}_{6}C$ nucleus contains <u>6</u> protons and <u>8</u> (14 - 6) neutrons. Therefore, the <u>calculated</u> mass of the carbon-14 nucleus is:

$$m_{calcd.} = (6 \times 1.008142 \text{ amu}) + (8 \times 1.008982 \text{ amu}) = 14.120708 \text{ amu}.$$

The <u>observed</u> mass of a carbon-14 nucleus is 14.007682 amu. Therefore:

$$\text{Mass Difference} = 14.120708 \text{ amu} - 14.007682 \text{ amu} = 0.113026 \text{ amu}.$$

$$\text{Mass Difference} = (0.113026 \text{ amu})(1.66 \times 10^{-24} \text{ g/amu}) = 1.88 \times 10^{-25} \text{ g}.$$

Finally, convert this value into energy units, using Einstein's equation and the fact that 1 Joule = 1 kg m^2/sec^2. The result is:

$$\text{Binding Energy} = (1.88 \times 10^{-25} \text{ g})(1 \text{ kg}/1000 \text{ g})(3.00 \times 10^8 \text{ m/sec})^2$$

$$\text{Binding Energy} = 1.69 \times 10^{-11} \text{ kg m}^2/\text{sec}^2 = \underline{1.69 \times 10^{-11} \text{ joules}}.$$

For comparison purposes, the <u>binding energy per nucleon</u> is used. (A "nucleon" is a proton or neutron.) In this case, (1.69×10^{-11} J)/14 = <u>1.21×10^{12} J per nucleon.</u>

---

1) Masses obtained from "<u>The Atomic Nucleus</u>" by R. D. Evans. 1955 by McGraw-Hill, Inc., page 137.

PARTICLE ACCELERATORS AND PARTICLE DETECTORS:

Transmutation of one element into another can be made to occur by bombarding an atomic nucleus with high-energy particles such as protons and electrons. For many years, the only way to obtain such high-energy particle beams was by using radioactive atoms as their source. It is now possible, however, to obtain beams of high-energy particles by using instruments known as particle accelerators. Particle accelerators use electric and magnetic fields to accelerate charged particles to high speeds. An example of a particle accelerator is the cyclotron, which uses magnetic fields to cause charged particles to move in circles. Linear particle accelerators are used if a straight-line pathway for the particles is desired -- the linear particle accelerator at Stanford University is about two miles long! Larger particle accelerators are projected for the future, among them the "superconducting super collider" which will be built underground in Texas.

The presence of high-energy particles and the pathways they travel can be detected using instruments known as particle detectors. Probably the best-known particle detector is the Geiger-Müller counter, which uses electronic methods to detect ionizing radiation. (Ionizing radiation is any form of radiation which strips electrons away from the atoms it encounters. Gamma radiation is one example of ionizing radiation.) The pathways traveled by particles can be detected using a cloud chamber, in which "trails" are left by particles which pass through the vapor from a volatile liquid. (This is similar to the "vapor trails" left by jet planes when they pass through the water vapor in the air.) Other particle detectors include the scintillation counter, which contains a substance such as zinc sulfide which glows when high-energy particles strike it (a "scintillation" is a spark or flash), and the dosimeter, which contains a piece of photographic film which gets darker when irradiated by gamma radiation. Nuclear chemists wear dosimeters daily.

MEASURING RADIOACTIVITY:

Radioactivity is measured in units called Becquerels. A radioactive sample which emits one particle per second has a radioactivity of one Becquerel. (The unit is named for the French scientist Henri Becquerel, who shared the 1903 Nobel Prize in Physics for his discovery of radioactivity.) A more convenient unit for most laboratory work is the Curie. One Curie = 3.7×10^{10} Becquerels, which is the radioactivity of 1.0 gram of radium. (The Curie is named in honor of Mme. Marie Curie, who shared the 1903 Nobel Prize in Physics with Becquerel and her husband Pierre. Mme. Curie also won the 1911 Nobel Prize in Chemistry for her discovery of the elements radium and polonium.)

Of greater interest are the units used to measure radiation dosages in living tissues. These are derived from a unit called the Roentgen, which is the amount of gamma radiation needed to ionize 2.08×10^9 air molecules per milliliter of air at STP. (The Roentgen is named for the German physicist Wilhelm Roentgen, who won the 1901 Nobel Prize in Physics for his discovery of X-rays, which were originally called "Roentgen Rays". X-rays consist of photons which generally have less energy than gamma rays. Their energy is measured in electron volts (eV). One electron volt is the amount of energy gained by an electron which passes through a potential difference of one volt. 1.00 eV = 1.60×10^{-19} joules.) The most useful of the derived units for radiation dosage is the rem, which stands for "Roentgen equivalent in man". One rem is the amount of radiation -- any kind, not just gamma radiation -- that produces the same biological effect in human tissues as would be produced by one roentgen of gamma radiation. The federal government has set a limit of 500 millirems per year as the maximum "safe" radiation dosage for the general public.[1] (Of course, the only completely safe dosage is zero, but this is unlikely.)

1) From "The Straight Dope" by Cecil Adams. 1984 by Chicago Reader, Incorporated, page 238.

EFFECTS OF RADIATION ON THE HUMAN BODY:

Each person on earth receives a certain dosage of radiation each year from natural sources. Much of this radiation is in the form of cosmic rays, which are largely made up of alpha, beta, and gamma particles which have their origin in nuclear reactions which are occurring in outer space. The gamma radiation is especially harmful because it is ionizing radiation -- it removes electrons from atoms, leaving positive ions behind. The problem is that if the lost electron was part of an electron pair, the resulting ion is a free radical. Free radicals have been shown to be probable cancer-causing agents. An average person in the United States receives about 100 millirems of "background radiation" each year from cosmic rays, X-rays, and other sources. (X-rays are produced by a process known as K-capture, in which an electron in the innermost shell of an atom (the "K" shell) is captured by the atomic nucleus, converting a proton into a neutron in the process: ${}^{1}_{1}p^{+} + {}^{0}_{-1}e^{-} \rightarrow {}^{1}_{0}n$. X-rays are used in medicine and dentistry to examine bones and teeth for structural irregularities, and in other ways as well.) The amount of "background radiation" varies with latitude, as the earth's magnetic field deflects a large amount of cosmic radiation away from the earth. People who travel in airplanes a great deal may receive more cosmic radiation than most, since they are not as shielded by the earth's magnetic field as those who remain on the earth's surface.[1] (The same is true for people who work in nuclear power plants!)

Radiation is used by the medical profession in certain beneficial ways. For example, iodine is absorbed by the thyroid gland, so radioactive iodine-131 can be given internally to people who have thyroid conditions as either a diagnostic or a therapeutic measure. Vitamin B_{12} contains cobalt ions, so a patient who has a Vitamin B_{12} deficiency may be fed some cobalt-60 so Vitamin B_{12} can be monitored.

---

1) From "The Straight Dope" by Cecil Adams. 1984 by Chicago Reader, Incorporated, page 238.

RADIOCARBON DATING:

Radiocarbon dating is a method used to determine the age of artifacts by measuring the amount of radioactive carbon-14 present in them. Carbon-14 is produced in the atmosphere when a free neutron collides with a nitrogen atom:

$$^{1}_{0}n + ^{14}_{7}N \longrightarrow ^{14}_{6}C + ^{1}_{1}H.$$

The carbon-14 is oxidized to CO_2 by oxygen in the atmosphere, and the CO_2 is taken in by plants, which in turn are eaten by animals. Thus, every living thing keeps on ingesting radioactive carbon-14 until it dies. After the death of an organism, the carbon-14 present undergoes radioactive decay with a half-life of 5730 years. (Radioactive decay processes follow first-order kinetics, and therefore have half-lives which are constants. This makes determining the age of artifacts very easy.) By comparing the amount of carbon-14 present in an artifact with the amount present in living tissues, the age of the artifact can be calculated, as shown below.

Problem[1]: The radioactivity of the carbon-14 in a skull fragment is 0.016 Becquerels/gram. The radioactivity of carbon-14 in living tissue is 0.255 Becquerels/gram. Calculate the age of the skull fragment.

Solution: After each half-life, the radioactivity of the fragment will be half of what it was originally. Thus, after one half-life, the radioactivity is 0.255/2 = 0.128 Becquerels/gram; after two half-lives, it is 0.128/2 = 0.064 Bq/g; and so on. Therefore, the problem can be represented by the following equation:

$$(0.016 \text{ Becquerels/gram}) = (0.255 \text{ Becquerels/gram})(1/2)^n.$$

Solving this equation for n (either logarithmic methods or "trial and error" can be used) gives a value of 4, which means that four half-lives have elapsed since the death of the organism. Each half-life is 5730 years, so:

Age of Skull Fragment = 4 x 5730 years = 22,920 years.

---

1) From a similar problem on page 939 in "Chemistry: An Experimental Science" by G. M. Bodner and H. L. Pardue. 1989 by John Wiley & Sons, Inc.

NUCLEAR ENERGY:

Nuclear energy can be generated by either of two processes: nuclear fission or nuclear fusion. Nuclear fission involves the splitting of one large atomic nucleus into smaller atomic nuclei. Consider the following nuclear reaction:

$$^{235}_{92}U + ^{1}_{0}n \rightarrow ^{94}_{36}Kr + ^{139}_{56}Ba + 3\ ^{1}_{0}n + \text{energy.}$$

The above equation shows that bombarding a uranium-235 nucleus with a neutron gives two smaller nuclei and three neutrons, in addition to giving off large amounts of energy. The three neutrons produced can each strike another uranium nucleus and cause the reaction to occur again, unless they are absorbed by some other substance such as cadmium. This is the basis for a nuclear "chain reaction", and is the basic process used in nuclear power plants. "Meltdowns" occur when too much energy or too many neutrons are produced in a short period of time. Nuclear reactors are now being designed in better ways so as to make meltdowns less likely.[1]

Nuclear fusion involves the joining of two or more small atomic nuclei to form one large atomic nucleus. Nuclear fusion occurs in stars, which give off large amounts of energy in the process of fusing hydrogen nuclei to form helium:

$$^{2}_{1}H + ^{3}_{1}H \rightarrow ^{4}_{2}He + ^{1}_{0}n + \text{energy.}$$

(In a sense, then, solar energy is nuclear energy, since sunlight is produced by nuclear processes.) The above reaction produces 17.0×10^6 electron volts of energy per helium atom produced, so fusion research is being actively pursued as a possible source of energy. (In March of 1989, scientists at the University of Utah announced that they had achieved nuclear fusion in a beaker of water, but subsequent investigations showed their claims to be largely without merit.[2]) The raw material -- hydrogen -- for the fusion process is readily available from sea water (H_2O), but it is extremely difficult to get two positively-charged nuclei close enough to fuse.

---

1) U. S. News & World Report, May 29, 1989, pages 52-53.
2) For a full report, see Newsweek, May 8, 1989, pages 48-56.

APPENDIX -- CHEMICAL NOMENCLATURE:

Chemical nomenclature is just the system by which chemical compounds are named. The system for ionic compounds is different from the system for covalent compounds. Both systems are described below.

Ionic compounds are named by either of two methods. In the first method, the cation is named first, and the anion is named with the suffix "-ide". Thus, KBr is potassium bromide, and CaO is calcium oxide. When transition metals (which form more than one cation) are used, the charge on the cation is included in Roman numerals after the cation's name. Thus, CuCl is copper(I) chloride, and $CuCl_2$ is copper(II) chloride. The other method that is sometimes used for naming transition metal compounds involves using the metal's Latin name, with a suffix attached to distinguish between the possible cations. Under this system, CuCl is cuprous chloride, whereas $CuCl_2$ is cupric chloride. (The suffix "-ic" indicates the cation with the larger charge, while the suffix "-ous" indicates the cation with the smaller charge.) It is important to keep the two systems separate: FeO is either iron(II) oxide or ferrous oxide, whereas Fe_2O_3 is either ferric oxide or iron(III) oxide, but not "ironic oxide"!

Covalent compounds are named as though they were ionic compounds, with the "cation" first and the "anion" last. Greek prefixes are used to indicate the number of each atom present in the compound. The prefixes used are "mono-" (1), "di-" (2), "tri-" (3), "tetra-" (4), "penta-" (5), "hexa-" (6), "hepta-" (7), "octa-" (8), "nona-" (9), "deca-" (10), and so on. Thus, CO_2 is carbon dioxide, CCl_4 is carbon tetrachloride, CO is carbon monoxide, and N_2O_5 is dinitrogen pentoxide. (Note that in the cases of "monoxide" and "pentoxide", the prefix has been shortened slightly so as to make pronunciation easier.) This system should not be used with ionic compounds -- Fe_2O_3 is not "diiron trioxide"!

APPENDIX -- NOMENCLATURE OF ACIDS AND THEIR SALTS:

The rules for naming binary acids and their salts are different from the rules for naming oxyacids and their salts. Binary acids (those composed of hydrogen and one other element) are named by placing the prefix "hydro-" and the suffix "-ic acid" around the name (sometimes abbreviated) of the "other" element. Thus, HCl is hydrochloric acid, and HI is hydriodic acid. The salts of binary acids have names which end in "-ide". For example, NaCl is sodium chloride, and KI is potassium iodide.

The names of oxyacids (those composed of hydrogen, oxygen, and one other element) are found using a somewhat more complex system of rules. Prefixes and suffixes are still used, but the particular combination of a prefix and a suffix is what provides information about the number of oxygen atoms present in the molecules of the acid. This situation arises because any given element may be able to form several oxyacids. If only two oxyacids are known for a given element, then prefixes are not used, and the suffixes "-ic acid" and "-ous acid" are used to indicate the larger number of oxygen atoms and the smaller number of oxygen atoms, respectively. Thus, HNO_3 is nitric acid, whereas HNO_2 is nitrous acid. If more than two oxyacids are possible for a given element, then the prefixes "hypo-" and "per-" are used to indicate the minimum number of oxygen atoms and the maximum number of oxygen atoms, respectively. Thus, $HClO_4$ is perchloric acid, $HClO_3$ is chloric acid, $HClO_2$ is chlorous acid, and HClO is hypochlorous acid. The salts of oxyacids whose names end in "-ic acid" have names which end in "-ate", while the salts of oxyacids whose names end in "-ous acid" have names which end in "-ite". Thus, KNO_3 is potassium nitrate, and NaClO is sodium hypochlorite. Salts which still contain acidic hydrogen atoms are named using a prefix to indicate the number of hydrogen atoms present. Thus, NaH_2PO_4 is sodium dihydrogen phosphate.

APPENDIX -- THE LAWS OF LOGARITHMS:

The logarithm of a number is the power to which ten (10) must be raised to obtain the number. The natural logarithm of a number is the power to which e (e = 2.71828...) must be raised to obtain the number. The abbreviation for "logarithm" is "log"; the abbreviation for "natural logarithm" is "ln". Some of the laws of logarithms with which you should be familiar are listed below.

Law of Logarithms	Example
$\log 10^a = a$	$\log 10{,}000 = \log 10^4 = 4.$
$10^{(\log a)} = a$	$10^{(\log 10{,}000)} = 10^4 = 10{,}000.$
$\log a^b = b \log a$	$\log 8 = \log 2^3 = 3 \log 2$
	$= 3 \times 0.301 = 0.903.$
$\log (a \times b) = \log a + \log b$	$\log 20 = \log (2 \times 10)$
	$= \log 2 + \log 10$
	$= 0.301 + 1 = 1.301.$
$\log (a/b) = \log a - \log b$	$\log 5 = \log (10/2) = \log 10 - \log 2$
	$= 1 - 0.301 = 0.699.$
$\log (1/a) = -\log a$	$\log 0.0001 = \log (1/10{,}000)$
	$= -\log 10{,}000 = -4.$
$\ln a = 2.303 \log a$	$\ln 100 = 2.303 \log 100$
	$= 2.303 \times 2 = 4.606.$
$\ln e^a = a$	$\ln 20.085 = \ln e^3 = 3.$
$e^{(\ln a)} = a$	$e^{(\ln 20.085)} = e^3 = 20.085.$

The other laws for natural logarithms are similar to the laws for the common logarithms (base = 10) above. Converting a number to its logarithm entails the "gain" of one significant digit. For example, $\log (2.0 \times 10^5) = 5.30$, since the "5" comes from the "10^5" and the ".30" comes from the "2.0" (two sig. figs.).